INTERNATIONAL CENTRE FOR MECHANICAL SCIENCES

COURSES AND LECTURES No. 238

STABILITY OF ELASTIC STRUCTURES

EDITED BY
H. LEIPHOLZ
UNIVERSITY OF WATERLOO

SPRINGER-VERLAG WIEN GMBH

Originally published by Springer - Verlag Wien New York in 1978

ISBN 978-3-211-81473-4 ISBN 978-3-7091-2975-3 (eBook)
DOI 10.1007/978-3-7091-2975-3

PREFACE

This monograph is based on the notes of a lecture series on Stability of Elastic Structures given at the International Centre for Mechanical Sciences (CISM) In Udine, Italy, from October 4 to October 13, 1976 by Professors H. H. E. Leipholz, K. Huseyin, both from the University of Waterloo, Ontario, Canada, and M. Zyczkowski from the Technical University of Cracow, Poland. The main objective of the lecture series was to report on the developments in the domain of the stability of elastic structures which have occurred during the last three decades, and to make the audience familiar with a number of new problems and methods for their solution as these are typical for the modern perception of stability.

Actually, there were new problems, posed to structural engineers in the age of nuclear technology and spacemechanics, that enforced the revision of the classical theory of stability of structures. The existence of deflection depending, so called follower forces had to be acknowledged, and it had to be admitted that the transition of a structure from a stable to an unstable state is essentially a dynamic process. Only in exceptional cases, which, however, were those mostly considered in Civil Engineering practice in the past, a static approach to stability problems is possible. In general, only the dynamic approach is powerful enough to reveal the stability behaviour of a structure to the full extent. This is the case, because the flutter phenomenon is not only restricted to structures subjected to time depending loading but may also be observed for structures subjected to follower loads which are time independent. In the light of these facts, it became clear that attempts had to be made to view the stability of elastic structures in the broader context of dynamical systems. As a consequence, methods developed in other areas of mechanics and engineering science had to be made suitable for an application to structures. This was specifically true for Liapunov's Second Method which is largely applied in Mechanical Engineering and Control Theory for the treatment of stability problems of discrete systems. However, structures are essentially continuous systems (if not approximately discretized). Therefore, Liapunov's method had to be reformulated and extended for an application to continuous systems. This specific subject will be treated thoroughly in this treatise.

The monograph consists of three parts. Each part is self-contained, but deep-rooted connections between the various parts can easily be detected. The first part deals with basic concepts, definitions, and criteria. The stability problem is mathematically formulated, and methods for the solution are presented and discussed. The second part emphasizes the multiple-parameter aspect of stability. Already by using the dynamic approach, two parameters, namely frequency and load, are involved. But, there may be several load parameters instead of a single one and in addition imperfection parameters may be contained in the problems concerned. As a consequence, the investigations are carried out in an n-dimensional space. Moreover, the algebraic method is being used, assuming that the problems are posed in terms of lumped-mass systems rather than in terms of continuous systems. The third part deals with some special and practical aspects of non-conservative systems subjected to follower forces which are of interest to the designer. Influence of damping, optimal design of columns and shells, and the investigation of the postbuckling behaviour of imperfect structures may be mentioned as some of the topics considered in that part of the monograph.

It is my pleasure to thank CISM for the support and hospitality which the three authors received during their stay at Udine. We owe special thanks to Professor W. Olszak, who suggested the course on Stability of Elastic

Structures and encouraged us to write this monograph. It is dedicated to all those devoted to problems of mechanics in connection with structural engineering and over all to those interested in that fascinating and practically so important subject of stability.

Waterloo, Cracow, Spring 1978

H. H. E. Leipholz

K. Huseyin

M. Życzkowski

CONTENTS

Page

PART I by H. H. E. Leipholz, University of Waterloo

1. Introduction 1

1.1 Basic concepts 1
1.2 Differential geometric aspects 4
1.3 Stability definitions, topological aspects 6

2. The Mathematical Formulation of the Stability Problem 9

2.1 Equation of motion 10
2.2 Variational equation, the fundamental problem 22

3. Approaches to the Solution 36

3.1 Liapunov's approach 36
3.2 Energy approach 61
3.3 Modal approach, descretization, algebraization 65

4. Conclusion 87

Figures 89

References 97

PART II by K. Huseyin, University of Waterloo

Introduction 100

1. Stability of Gradient Systems 100

1.1 Introductory remarks 100
1.2 Classification of critical conditions 102
1.3 Equilibrium surface 106
1.4 Stability boundary 109
1.5 Geometry of the equilibrium surface and connection with the catastrophe theory 112
1.6 Examples 116

2. Stability of Autonomous Systems 121

2.1 Definitions 121
2.2 Classification of systems 126
2.3 Conservative systems 128
2.4 Pseudo-conservative systems 134
2.5 Gyroscopic systems 141
2.6 Circulatory systems 156

Figures 161

References 179

Page

PART III by M. Zyczkowski, Technical University of Cracow, Poland

1. Influence of the Behaviour of Loading on Its Critical Value 181

1.1 Introduction 181
1.2 Derivation of the equation determining the exact value of the critical force 183
1.3 Analysis of the stability curve 189

Figures 198

References 200

2. Influence of Simultaneous Internal and External Damping on the Stability of Non-Conservative Systems 201

2.1 Introduction 201
2.2 Analysis of Ziegler's model 201
2.3 Particular cases 206

Figures 207

References 209

3. Interaction Curves in Non-Conservative Problems of Elastic Stability 211

3.1 Introduction 211
3.2 Assumptions 212
3.3 Static criterion of stability 213
3.4 Energy method 215

Figures 222

References 224

4. Optimal Design of Elastic Columns Subject to the General Conservative and Non-Conservative Behaviour of Loading 225

4.1 Introduction 225
4.2 Statement of the problem in the general conservative case 225
4.3 General considerations 229
4.4 Particular solutions 233
4.5 Optimization of an elastic bar compressed by antitangential forces 236

Figures 238

References 241

Page

5. Investigation of Postbuckling Behaviour of Imperfect Cylindrical Shells by Means of Generalized Power Series 243

5.1 Introduction 243
5.2 Non-linear problem of stability of a circular cylindrical shell subject to hydrostatic loading 244
5.3 Elimination of the parameter d for moderate-length shells 247
5.4 Pressure in terms of the deflection amplitude 251
5.5 Upper critical pressure 252
5.6 Lower critical pressure 258
5.7 Summary of results written in physical quantities 263

References 265

6. Optimal Design of Shells with Respect to Their Elastic Stability 268

6.1 Introduction 268
6.2 Parametric optimal design of a spherical panel 269
6.3 Optimal design of a cylindrical shell under pure bending 272

Figures 285

References 290

PART I

by H.H.E. Leipholz
University of Waterloo
Ontario Canada

1. Introduction

Concepts and quantities used in stability theory are to a large extent not invariant. They are chosen and defined according to the particular intent of the researcher and the purpose of his investigation. In many cases, practical aspects of the problems involved dictate the point of view to be adopted for the approach to stability. Therefore, a stability theory as such does, strictly speaking, not exist. It is necessary, before starting any stability considerations, to define clearly the stability concepts to be used in order to avoid misunderstanding and confusion. According to the chosen concepts and definitions, the specific stability theory for a specific situation is then developed. We shall proceed according to these guidelines in the following.

1.1 Basic concepts

The behaviour of an elastic system is described, by one or more characteristics which exhibit at the onset of instability a certain property suitable for the formulation of a stability criterion. In a general state, the system possesses a degree of stability which is the norm of the perturbation (magnitude) necessary to drive the system to the stability boundary and slightly beyond it. In addition, the system depends on a variety of parameters, e.g. load parameters, structural parameters and possibly the time, which may act as a parameter by, for example, effecting the elastic properties of the system through aging. The parameters control the system's behaviour as the degree of stability depends on the parameter values. At critical values, the degree of stability vanishes, and a perturbation of arbitrary smallness can destabilize the system. The aim of a stability investigation is to determine the critical parameter values.

In order to illustrate the aforementioned concepts, the simple example shown in Figure 1 shall be considered. A ball of mass m subjected to gravity is located at point C in a crater whose depth is h. The edge AB of the crater is h feet above the bottom of the crater. Quantity h is supposed to be variable and constitutes the control parameter of the problem as will be recognized soon. The actual distance r of the ball from its equilibrium position C shall serve as the characteristic of the system. The equilibrium position, whose stability is to be investigated, is therefore defined by the value $r_o \equiv 0$ of the characteristic r. If the ball is removed from its position of equilibrium C by a perturbation I = mv (which is an impulse), the characteristic changes from $r_o \equiv 0$ to $r = r_I > 0$. As long as the magnitude of I if small enough, the ball remains in the crater and is forced back to position C by gravity. That means, that the value r_o of the characteristic r is a stable one. If the magnitude of I is sufficiently large, the ball is forced to leave the crater. As a consequence, it will not return to position C but remain at a distance from C which may be finite or infinite. Position C, or value r_o of r, has become unstable. The magnitude of I just needed to cause instability may be called the degree of stability. Obviously, the degree of stability of C, or r_o, is governed by the magnitude of h. That explains why h shall be called the control paramater.

If h is allowed to tend towards zero, the impulse able to drive the ball out of the crater can be smaller and smaller, i.e., the degree of stability decreases. For h = 0, already an infinitesimal impulse I removes the ball from position C to which it never returns. Hence, the degree of stability has vanished, and the limit of stability has been reached. Since the degree of stability has become zero for $h = 0$, one can conclude that $h_{crit} = 0$ is the critical value of the control parameter h.

For $h < h_{crit}$, the stability limit of C, or r_o, is exceeded and instability prevails. This situation is shown in Figure 1c.

Summarizing, the following can be stated: investigating the stability of a mechanical system, a characteristic, e.g. r, must be chosen. In the most general case, the characteristic is a manifold in a paramater space. The problem to be solved consists in investigating the stability of certain prescribed points of this manifold. Stability may be impaired by a perturbation I effecting the characteristic r. Stability will be lost when the perturbation is strong enough to overcome the degree of stability. Then, the stability boundary is exceeded, and instability sets in. The occurrence of instability is indicated by a criterion based on a stability definition.

The stability definition may read as follows:

"Let

$$r = r(I), \quad r_o = r(0) \tag{1.1}$$

hold true. If in the neighbourhood of r_o, the size of which is determined by the degree of stability,

the condition

$$\left|\frac{dr}{dI}\right| = c, \quad c = \text{const.} < \infty \tag{1.2}$$

is satisfied for any value of r and I, and if in addition r tends towards r_o when I tends towards zero, then r_o is stable."

According to this definition,

$$\left|\frac{dr}{dI}\right| = \infty \tag{1.3}$$

is an instability criterion.

The stability of the mechanical system is governed by a control parameter, e.g. h. This parameter affects the degree of stability and can cause it to vanish completely. The value of the control parameter for which the degree of stability becomes equal to nought is called the critical value. This value is the aim of the investigation. As for parameter values beyond the critical one the system becomes unstable under the slightest perturbation, it is essential to know the critical parameter value for a proper design and operation of the system.

Further detailed considerations and examples can be found in [1] and [2].

1.2 Differential geometric aspects

Viewing the characteristic r as a hypersurface in the parameter space, as shown in Figure 2 for two parameters only, leads to a differential geometric interpretation of the stability definition. In the most elementary sense, the concept of stability comes into play if one observes the change of a quantity, say r, due to the change of a parameter, say P_1 or P_2. One tends to say, that the quantity r behaves in a stable way at a certain point P_i^*, if the increment of r is finite for a finite increment of P_i, $i = 1,2$, and if the increment of r is a continuous function of the increment of P_i so that dr tends towards zero if dP_i so does. Analytically, it means that the r-surface of P_i^* is continuous and possesses well-defined, finite differential quotients. A consideration like this one is a justification of the previously introduced stability definition and the instability criterion (1.3). Now the first, crude perception of stability as given in section 1.1 shall be refined.

Assuming the characteristic r dependent on several parameters leads, as already mentioned, to the hypersurface concept. Agreeing on the fact that it is the smoothness of the r-surface around a certain point P_i^* that indicates stability at that point, one may investigate the quantity

$$dr = [\text{grad } r]_{P_i^*} \cdot d\underline{u}, \tag{1.4}$$

where

$$d\underline{u} = \alpha_i \underline{b}_i, \tag{1.5}$$

α_i are direction cosines, and $\underline{b}_i$ the unit vectors along the axes of the P_i space. Expression (1.4) is the directional derivative of r in the chosen direction $d\underline{u}$ at P_i^*. In generalizing (1.2), one may say that r is stable around P_i^* if $|dr|$ as given by (1.4) is finite. Since $d\underline{u}$ is arbitrary, it is therefore the quantity $[\text{grad } r]_{P_i^*}$ which decides on stability. Hence, the stability investigation assumes a differential geometric aspect: at any point of the r-surface where grad r is either zero or not finite or not defined at all, one must expect the occurrence of instability.

Another quantity that belongs to the intrinsic geometry of the r-surface is its Gaussian curvature K. It is known that

$$K = \frac{L}{g}, \quad L = \det(L_{ik}), \quad g = \det(g_{ik}),$$

where L_{ik} are the coefficients of the second fundamental form of the surface and g_{ik} are the metric coefficients. However,

$$L_{ik} = \underline{n} \cdot \frac{\partial^2 \underline{R}}{\partial P_i \partial P_k}$$

where $\underline{n}$ is the unit normal to the surface and $\underline{R}$ the position vector leading from the origin of the parameter space to the point on the surface in question.

Since

$$\text{grad } r = \lambda \underline{n},$$

where λ is a proportionality factor, also

$$L_{ik} = \frac{1}{\lambda} \text{ grad } r \cdot \frac{\partial^2 \underline{R}}{\partial P_i \partial P_k} .$$

Hence, it becomes obvious that K, the Gaussian curvature, depends on grad r through L_{ik}. Therefore, it has to be expected that at points of the r-surface where instability occurs not only grad r behaves irregularily but also K. The conclusion is that instability can also be detected by studying the Gaussian curvature and determining the points of the r-surface where K is zero or infinite. An example for a stability investigation using the Gaussian curvature, can be found in [3].

1.3 Stability definitions, topological aspects

Let again the r-surface shown in Figure 2 be considered. As already mentioned earlier, the stability around a certain point of the r-surface depends on the smoothness conditions encountered around that point. These conditions are expressible in terms of differential quotients, and since the directions in which these quotients can be calculated are numerous, there are also numerous stability conditions and definitions which may be taken into account. In Figure 2, some of them are listed.

The most elementary case is the one, where only the smoothness around a particular point (P_1^*, P_2^*) of the r-surface is being considered. Then, one deals with local stability with respect to changes of P_i, i = 1,2. The analytical condition for local stability is

$$\left| \frac{\partial r}{\partial P_i} \right| = c < \infty, \quad i = 1,2, \text{ at } (P_1^*, P_2^*). \tag{1.6}$$

The most general case is that of global stability. Then, the r-surface is smooth at any point within a certain limited or unlimited domain. The corresponding analytical condition for global stability reads

$$\left|\frac{\partial r}{\partial P_i}\right| = c < \infty, \quad i = 1,2, \quad \text{at any point } (P_1,P_2). \tag{1.7}$$

Sometimes, one encounters a sitatuion between local and global stability. One may then use the term semi-global stability. It is, for example, stability along a trajectory $P_2 = P_2^*$, and it can be stability with respect to any parameter changes along this trajectory. A possible analytical condition for semi-global stability is

$$\left|\frac{\partial r}{\partial P_1}\right| = c < \infty \text{ at } P_2 = P_2^* \text{ and any value of } P_1. \tag{1.8}$$

Another analytical condition for semi-global stability reads

$$\left|\frac{\partial r}{\partial P_2}\right| = c < \infty \text{ at } P_2 = P_2^* \text{ and any value of } P_1. \tag{1.9}$$

Condition (1.9) is a case of great practical importance and may be called intensified Liapunov stability: The justification of this denotation will be given soon.

It may be pointed out that all the preceding statements can be generalized to cover the case where a smooth and analytical r-surface does not exist. Then, the definitions (1.6) to (1.9) may be reformulated in terms of ratios of increments of the quantities involved. That may be necessary, as differential quotients may not exist. For example, instead of (1.9), one may introduce the original Liapunov stability definition which reads "r is Liapunov stable at $(0,P_2^*)$ if for positive quantities ε and $\eta = \eta(\varepsilon)$ the condition

$$|r(P_1,P_2^*+dP_2) - r(P_1,P_2^*)| \leqq \varepsilon \text{ at } P_1 \geqq 0$$

follows from the condition

$$|r(0,P_2^*+dP_2) = r(0,P_2^*)| \leqq \eta \text{ at } P_1 = 0 \qquad \text{(LD)}$$

for arbitrary small ε."

Obviously, definition (LD) includes (1.9) which is a more restricted version of the Liapunov stability definition. Therefore, speaking of intensified Liapunov stability [4], may be reasonable. This can be seen as follows: condition (1.9) can be rewritten to yield

$$|r(P_1,P_2^*+dP_2) - r(P_1,P_2^*)| \leqq |dP_2| .c$$

This inequality is supposed to hold for any value of P_1. Also choosing $|dP_2|$ so small that $|dP_2|.c = \varepsilon$, where ε is arbitrary small, the previous condition can be changed to read

$$|r(P_1,P_2^*+dP_2) - r(P_1,P_2^*)| \leqq \varepsilon \text{ at } P_1 \geqq 0.$$

This version coincides already with the first condition of (LD). Moreover, (1.9) holds at any value of P_1, hence, also for $P_1 \equiv 0$. Therefore

$$|r(0,P_2^*+dP_2) - r(0,P_2^*)| \leqq |dP_2| .c = \varepsilon \text{ at } P_1 = 0.$$

But this condition is not in contradiction with the second condition of (LD). If for example the (<)-sign holds, then an $\eta < |dP_2|.c = \varepsilon$ can be found so that the second condition of (LD) is exactly satisfied. If, however, the (=)-sign holds, then one has to assume $\eta = \varepsilon$ in order to be again in line with the second condition of (LD). Obviously, condition (1.9) agrees completely with Liapunov's stability definition (LD).

The conclusion to be drawn is that depending on the situation encountered on the r-surface, as well as depending on the conditions imposed by the specific problem in question, for example with regard to the nature of the perturbation, i.e. the kind of parameter to be changed, a number of different stability definitions and conditions may be taken into account. One of those is Liapunov's definition in its original form (LD) or in its increased version (1.9). It is a very common and useful definition and will be used frequently in the following. However, one should keep in mind, that it may be replcaed by a weaker one like (1.6), if needed, from case to case.

For condition (1.9) it may now be indicated how instability can assume a strong topological aspect. For this purpose use the mapping shown in Figure 3. Let all points (P_1, P_2^*) for any value of P_1 coincide with the centre 0. Let the quantity $|P_1|$ be interpreted as an angle and the quantity $|dr|$ following from (1.9) for this specific value of P_1 be ploted as a radius vector from 0 under this angle. Then the tips of these radius vectors prescribe a curve which is completely contained in the disc having the radius $\varepsilon = |dP_2|c$, if condition (1.9) is satisfied. In this way, the stability problem has in fact been reduced to a more topological problem: In order to ensure stability, one has to show that all image points (tips of the radius vectors) remain in the neighbourhood of the centre 0 limited by ε. But that is a topological problem, as topology is the "science of position".

2. The Mathematical Formulation of the Stability Problem

After having chosen the characteristic r of the problem according to selfimposed conditions following from the specific purpose of the intended investigation an expression must be found which relates r with the parameters P_i of the system in question, and with the time t. This will usually be a differential equation in which r appears as the dependent variable, the P_i as

$$\int_V \delta \mathscr{U} dV = \int_V \left(\frac{\partial \mathscr{U}}{\partial w} \delta w + \frac{\partial \mathscr{U}}{\partial w_{x_1}} \delta w_{x_1} + \frac{\partial \mathscr{U}}{\partial w_{x_2}} \delta w_{x_2} + \frac{\partial \mathscr{U}}{\partial w_{x_1 x_1}} \delta w_{x_1 x_1} \right.$$

$$\left. + \frac{\partial \mathscr{U}}{\partial w_{x_2 x_2}} \delta w_{x_2 x_2} + \frac{\partial \mathscr{U}}{\partial w_{x_1 x_2}} \delta w_{x_1 x_2} \right) dV, \tag{2.4}$$

which is part of (2.1), shall now be rewritten. In (2.4) it has been assumed that $\mathscr{U}$ depends on w and its derivatives

$$w_{x_i} = \frac{\partial w}{\partial x_i}, \quad i = 1,2, \quad w_{x_i x_j} = \frac{\partial^2 w}{\partial x_i x_j}, \quad i,j = 1,2, \tag{2.5}$$

where x_i, $i = 1,2$, are the spatial coordinates. For a transformation of (2.4), the formula

$$\int_V u \frac{\partial v}{\partial x_i} dV = - \int_V v \frac{\partial u}{\partial x_i} dV + \int_B uv \cos (n,x_i) dB \tag{2.6}$$

shall be used. In (2.6), u and v are arbitrary functions of sufficient smoothness, dB is oriented as shown in Figure 4, and therefore provided with an algebraic sign, and

$$\cos (n,x_1) = \cos \alpha, \quad \cos (n,x_2) = \sin \alpha, \tag{2.7}$$

where the angle α is given in Figure 4.

Using

$$u = \frac{\partial \mathscr{U}}{\partial w_{x_1}}, \quad v = \delta w$$

in (2.6) and taking

$$\delta w_{x_1} = \delta \frac{\partial w}{\partial x_1} = \frac{\partial}{\partial x_1} \delta w$$

into account yields

boundary conditions. Initial conditions may be given explicitely, while differential equation and boundary conditions follow from an extended version of Hamilton's principle, called in [5] the extended, fundamental, variational principle, This principle reads in terms of the deflection w

$$\int_T \Bigg[\int_V \Big\{ \frac{d}{dt}\Big(\frac{\partial \mathcal{T}}{\partial w_t}\Big)\delta w + \delta \mathcal{U} - (Q[w] + F(\underset{\sim}{x},t))\delta w \Big\} dV$$

$$- \int_{B_S} \Big[\underline{\mathcal{F}^*}(w)\,\underline{\mathcal{D}^*}(\delta w) + \underline{\mathcal{F}^*}(\delta w)\,\underline{\mathcal{D}^*}(w) \Big] dB - \int_B P[w]\delta w dB \Bigg] dt = 0. \quad (2.1)$$

In (2.1), the following notations have been used: t is the time, T is a certain time interval, d/dt stands for differentiation with respect to time, $w_t = \partial w/\partial t$, V is the volume of the system, B is its boundary which consists of a supported part B_S and a free part B_F, $\mathcal{T}$ is the density of the kinetic energy

$$T = \int_V \mathcal{T} dV, \quad (2.2)$$

$\mathcal{U}$ is the density of the potential energy

$$U = \int_V \mathcal{U} dV, \quad (2.3)$$

δw is a virtual deflection, Q[w] stands for all forces not possessing a potential distributed over the volume of the system, P[w] comprises all forces without a potential distributed over the boundary of the system, $F(\underset{\sim}{x},t)$ is the prescribed driving force, $\underset{\sim}{x}$ is the "vector" of the spatial coordinates, $\underline{\mathcal{F}^*}$ is the "vector" of those internal forces which become apparent when the boundary B_S is released from its constraints, $\underline{\mathcal{D}^*}$ is the vector of the corresponding deflections.

The integral

constant or time dependent coefficients, and t as one of the independent variables which also involve spatial coordinates. Finally, a forcing function will be included which depends on t as well as on the spatial coordinates.

If this differential equation could be integrated, the r-surface discussed in the previous section were obtained and could be used for the discussion of stability. Usually, it is however assumed that this integration is not being performed at all or at least not completely. One may try to use directly the differential equation for an evaluation of stability. Such an approach is, for example, based on Liapunov's direct method. Or one may work with intermediate integrals only, avoiding the complete integration of the differential equation. A technique of this kind is the foundation of the energy approach. Finally, one may assume that the solutions is of a certain structure involving a quantity whose behaviour decides on stability or instability. One may then again avoid determining the solution explicitely, restricting oneself to a discussion of that quantity as a function of the control parameters. The so called model approach provides the frequency ω as such a quantity. All these possible approaches will be discussed in detail later on.

Regardless which route might be chosen, the basis of any further consideration is the differential equation of the characteristic. If for example the characteristic is the predominant deflection w of the system (perpendicular to the middle surface for a plate etc.), one may call its differential equation the equation of motion. The following section will be devoted to the question how to derive the equation of motion systematically.

2.1 Equation of motion

The behaviour of the elastic system under investigation is completely determined by a differential equation and the corresponding initial and

$$\int_V \frac{\partial \mathcal{U}}{\partial w_{x_1}} \delta w_{x_1} \, dV = - \int_V \frac{\partial}{\partial x_1} \left(\frac{\partial \mathcal{U}}{\partial w_{x_1}} \right) \delta w dV + \int_B \frac{\partial \mathcal{U}}{\partial w_{x_1}} \delta w \cos \alpha dB. \tag{2.8}$$

Correspondingly,

$$\int_V \frac{\partial \mathcal{U}}{\partial w_{x_1}} \delta w_{x_2} \, dV = - \int_V \frac{\partial}{\partial x_2} \left(\frac{\partial \mathcal{U}}{\partial w_{x_2}} \right) \delta w dV + \int_B \frac{\partial \mathcal{U}}{\partial w_{x_2}} \delta w \sin \alpha dB. \tag{2.9}$$

is obtained.

Now,

$$u = \frac{\partial \mathcal{U}}{\partial w_{x_1 x_1}}, \quad v = \delta w_{x_1}$$

and

$$\delta w_{x_1 x_1} = \frac{\partial}{\partial x_1} \delta w_{x_1}$$

shall be used in (2.6). The result is

$$\int_V \frac{\partial \mathcal{U}}{\partial w_{x_1 x_1}} \delta w_{x_1 x_1} \, dV = - \int_V \frac{\partial}{\partial x_1} \left(\frac{\partial \mathcal{U}}{\partial w_{x_1 x_1}} \right) \delta w_{x_1} \, dV + \int_B \frac{\partial \mathcal{U}}{\partial w_{x_1 x_1}} \delta w_{x_1} \cos \alpha dB. \tag{2.10}$$

The first term on the right hand side of (2.10) can again be rewritten by means of (2.6). For this purpose set

$$u = \frac{\partial}{\partial x_1} \left(\frac{\partial \mathcal{U}}{\partial w_{x_1 x_1}} \right), \quad v = \delta w, \quad \delta w_{x_1} = \frac{\partial}{\partial x_1} \delta w$$

which leads to

$$- \int_V \frac{\partial}{\partial x_1} \left(\frac{\partial \mathcal{U}}{\partial w_{x_1 x_1}} \right) \delta w_{x_1} \, dV = \int_V \frac{\partial^2}{\partial x_1^2} \left(\frac{\partial \mathcal{U}}{\partial w_{x_1 x_1}} \right) \delta w dV - \int_B \frac{\partial}{\partial x_1} \left(\frac{\partial \mathcal{U}}{\partial w_{x_1 x_1}} \right) \delta w \cos \alpha dB. \tag{2.11}$$

Using (2.11) in (2.10) yields

$$\int_V \frac{\partial \mathcal{U}}{\partial w_{x_1x_1}} \delta w_{x_1x_1} dV = \int_V \frac{\partial^2}{\partial x_1^2}\left(\frac{\partial \mathcal{U}}{\partial w_{x_1x_1}}\right) \delta w dV - \int_B \frac{\partial}{\partial x_1}\left(\frac{\partial \mathcal{U}}{\partial w_{x_1x_1}}\right) \delta w \cos \alpha dB$$

$$+ \int_B \frac{\partial \mathcal{U}}{\partial w_{x_1x_1}} \delta w_{x_1} \cos \alpha dB \,. \tag{2.12}$$

But,

$$\delta w_x = \frac{\partial}{\partial x_1} (\delta w) = - \frac{\partial}{\partial B} (\delta w) \sin \alpha + \frac{\partial}{\partial n} (\delta w) \cos \alpha, \tag{2.13}$$

where $\partial/\partial B$ denotes differentiation along B and $\partial/\partial n$ differentiation along the normal n to B. Therefore, the last term in (2.12) changes to

$$\int_B \frac{\partial \mathcal{U}}{\partial w_{x_1x_1}} \delta w_{x_1} \cos \alpha dB = - \int_B \frac{\partial \mathcal{U}}{\partial w_{x_1x_1}} \frac{\partial}{\partial B} (\delta w) \sin \alpha \cos \alpha dB$$

$$+ \int_B \frac{\partial \mathcal{U}}{\partial w_{x_1x_1}} \frac{\partial}{\partial n} (\delta w) \cos^2 \alpha dB. \tag{2.14}$$

Integrating the first term in (2.14) by parts and using in the second term the identity

$$\frac{\partial}{\partial n} (\delta w) = \delta w_n$$

yields

$$\int_B \frac{\partial \mathcal{U}}{\partial w_{x_1x_1}} \delta w_{x_1} \cos \alpha dB = \int_B \frac{\partial}{\partial B}\left(\frac{\partial \mathcal{U}}{\partial w_{x_1x_1}}\right) \delta w \sin \alpha \cos^2 \alpha dB$$

$$+ \int_B \frac{\partial \mathcal{U}}{\partial w_{x_1x_1}} \delta w_n \cos^2 \alpha dB. \tag{2.15}$$

Using (2.15) in (2.12) yields finally

$$\int_V \frac{\partial \mathcal{U}}{\partial w_{x_1x_1}} \delta w_{x_1x_1} \, dV = \int_V \frac{\partial^2}{\partial x_1^2}\left(\frac{\partial \mathcal{U}}{\partial w_{x_1x_1}}\right) \delta w \, dV - \int_B \frac{\partial}{\partial x_1}\left(\frac{\partial \mathcal{U}}{\partial w_{x_1x_1}}\right)\delta w \cos \alpha dB$$

$$+ \int_B \frac{\partial}{\partial B}\left(\frac{\partial \mathcal{U}}{\partial w_{x_1x_1}}\right) \delta w \sin \alpha \cos \alpha dB + \int_B \frac{\partial \mathcal{U}}{\partial w_{x_1x_1}} \delta w_n \cos^2 \alpha dB. \tag{2.16}$$

By means of (2.6), (2.7), and of

$$\delta w_{x_2} = \frac{\partial}{\partial x_2} (\delta w) = \frac{\partial}{\partial B} (\delta w) \cos \alpha + \frac{\partial}{\partial n} (\delta w) \sin \alpha, \tag{2.17}$$

the expression

$$\int_V \frac{\partial \mathcal{U}}{\partial w_{x_2x_2}} \delta w_{x_2x_2} \, dV = \int_V \frac{\partial^2}{\partial x_2^2}\left(\frac{\partial \mathcal{U}}{\partial w_{x_2x_2}}\right)\delta w dV - \int_B \frac{\partial}{\partial x_2}\left(\frac{\partial \mathcal{U}}{\partial w_{x_2x_2}}\right)\delta w \sin \alpha dB$$

$$- \int_B \frac{\partial}{\partial B}\left(\frac{\partial \mathcal{U}}{\partial w_{x_2x_2}}\right) \delta w \sin \alpha \cos \alpha dB + \int_B \frac{\partial \mathcal{U}}{\partial w_{x_2x_2}} \delta w_n \sin^2 \alpha dB \tag{2.18}$$

can be derived in a corresponding way.

Also (2.6), (2.7), (2.13), and (2.17) can be used to derive similarly the relationship

$$\int_V \frac{\partial \mathcal{U}}{\partial w_{x_1x_2}} \delta w_{x_1x_2} \, dV = \int_V \frac{\partial^2}{\partial x_1 \partial x_2}\left(\frac{\partial \mathcal{U}}{\partial w_{x_1x_2}}\right)\delta w dV - \frac{1}{2}\int_B \frac{\partial}{\partial x_2}\left(\frac{\partial \mathcal{U}}{\partial w_{x_1x_2}}\right)\delta w \cos \alpha dB$$

$$- \frac{1}{2}\int_B \frac{\partial}{\partial x_1}\left(\frac{\partial \mathcal{U}}{\partial w_{x_1x_2}}\right)\delta w \sin \alpha dB + \frac{1}{2}\int_B \frac{\partial}{\partial B}\left(\frac{\partial \mathcal{U}}{\partial w_{x_1x_2}}\right)\delta w \sin^2 \alpha dB$$

$$- \frac{1}{2}\int_B \frac{\partial}{\partial B}\left(\frac{\partial \mathcal{U}}{\partial w_{x_1x_2}}\right)\delta w \cos^2 \alpha dB + \int_B \frac{\partial \mathcal{U}}{\partial w_{x_1x_2}} \delta w_n \sin \alpha \cos \alpha dB. \tag{2.19}$$

Let the following expressions and quantities be introduced: the variational derivative of U, that is to say,

$$\frac{\delta U}{\delta w} = \frac{\partial \mathcal{U}}{\partial w} - \frac{\partial}{\partial x_1}\left(\frac{\partial \mathcal{U}}{\partial w_{x_1}}\right) - \frac{\partial}{\partial x_2}\left(\frac{\partial \mathcal{U}}{\partial w_{x_2}}\right) + \frac{\partial^2}{\partial x_1^2}\left(\frac{\partial \mathcal{U}}{\partial w_{x_1x_1}}\right)$$

$$+ \frac{\partial^2}{\partial x_2^2}\left(\frac{\partial \mathcal{U}}{\partial w_{x_2x_2}}\right) + \frac{\partial^2}{\partial x_1 \partial x_2}\left(\frac{\partial \mathcal{U}}{\partial w_{x_1x_2}}\right), \tag{2.20}$$

and the "vectors"

$$\mathcal{F} = (\Phi_1, \Phi_2), \quad \mathcal{D} = (w, w_n), \tag{2.21}$$

where

$$\Phi_1 = \frac{\partial \mathcal{U}}{\partial w_{x_1}} \cos\alpha + \frac{\partial \mathcal{U}}{\partial w_{x_2}} \sin\alpha - \frac{\partial}{\partial x_1}\left(\frac{\partial \mathcal{U}}{\partial w_{x_1x_1}}\right)\cos\alpha - \frac{\partial}{\partial x_2}\left(\frac{\partial \mathcal{U}}{\partial w_{x_2x_2}}\right)\sin\alpha$$

$$- \frac{1}{2}\left[\frac{\partial}{\partial x_1}\left(\frac{\partial \mathcal{U}}{\partial w_{x_1x_2}}\right)\sin\alpha + \frac{\partial}{\partial x_2}\left(\frac{\partial \mathcal{U}}{\partial w_{x_1x_2}}\right)\cos\alpha\right] + \frac{\partial}{\partial B}\left[\left(\frac{\partial \mathcal{U}}{\partial w_{x_1x_1}} - \frac{\partial \mathcal{U}}{\partial w_{x_2x_2}}\right)\sin\alpha \cos\right.$$

$$\left. + \frac{1}{2}\frac{\partial \mathcal{U}}{\partial w_{x_1x_2}}\sin^2\alpha - \frac{1}{2}\frac{\partial \mathcal{U}}{\partial w_{x_1x_2}}\cos^2\alpha\right], \tag{2.22}$$

$$\Phi_2 = \frac{\partial \mathcal{U}}{\partial w_{x\,x}}\cos^2\alpha + \frac{\partial \mathcal{U}}{\partial w_{x_2x_2}}\sin^2\alpha + \frac{\partial \mathcal{U}}{\partial w_{x_1x_2}}\sin\alpha\cos\alpha. \tag{2.23}$$

By means of (2.8), (2.9), (2.16), (2.18), (2.19), (2.20), (2.21), 2.22), and (2.23) the integral (2.4) can be rewritten to read

$$\int_V \delta\mathcal{U}\,dV = \int_V \frac{\delta U}{\delta w}\,\delta w\,dV + \int_B \mathcal{F}(w)\,\mathcal{D}(\delta w)\,dB. \tag{2.24}$$

It is worth mentioning that $\underset{\sim}{\mathcal{F}}$ is the "vector" of all possible internal forces and $\underset{\sim}{\mathcal{D}}$ the "vector" of all the corresponding deflections.

Substituting (2.24) back into (2.1) and using the fact that $B = B_S + B_F$, yields the <u>derived version</u> of the extended fundamental variational principle, namely,

$$\int_T \Bigg[\int_V \left\{ \frac{d}{dt}\left(\frac{\partial \mathcal{T}}{\partial w_t}\right) + \frac{\delta U}{\delta w} - (Q[w] + F(\underset{\sim}{x},t))\delta w \right\} dV$$

$$+ \int_{B_S} [\underset{\sim}{\mathcal{F}}(w)\,\underset{\sim}{\mathcal{D}}(\delta w) - \underset{\sim}{\mathcal{F}}{}^*(w)\underset{\sim}{\mathcal{D}}{}^*(\delta w) - \underset{\sim}{\mathcal{F}}{}^*(\delta w)\underset{\sim}{\mathcal{D}}{}^*(w) - P[w]\delta w]dB$$

$$+ \int_{B_F} [\underset{\sim}{\mathcal{F}}(w)\,\underset{\sim}{\mathcal{D}}(\delta w) - P[w]\delta w]dB \Bigg] dt = 0. \qquad (2.25)$$

This principle can only hold true for an arbitrary interval of time T if the integral over the volume V as well as the integrals over B_S and B_F vanish identically. Since the variation δw is arbitrary over the volume V, the volume integral in (2.25) can only vanish if the integrand vanishes. Hence,

$$\frac{d}{dt}\left(\frac{\partial \mathcal{T}}{\partial w_t}\right) + \frac{\delta U}{\delta w} - Q[w] - F(\underset{\sim}{x},t) = 0 \qquad (2.26)$$

must be satisfied. Equation (2.26) is the equation of motion of the elastic system. The boundary conditions are obtained by enforcing the integrals over B_S and B_F to vanish.

As an illustration, consider the elastic, rectangular plate shown in Figure 5. The edges $x_1 = 0$, $x_2 = b$ are simply supported. The edge $x_2 = 0$ is clamped. The edge $x_1 = a$ is free. The plate is subjected to follower forces g distributed over the plate's surface and to follower forces N distributed along two edges opposite to each other one of which is simply supported while the other one is free.

Obviously,

$$\mathcal{T} = \frac{1}{2} \mu w_t , \tag{2.27}$$

where μ is the mass density per unit surface. μ is assumed to be constant. Using D as the flexural rigidity of the plate and ν as Poisson's ratio,

$$\mathcal{U} = \frac{D}{2} \Big\{ (w_{x_1x_1} + w_{x_2x_2})^2 + 2(1 - \nu)(w_{x_1x_2}^{\;2} - w_{x_1x_1} w_{x_2x_2})$$

$$- \left[\frac{g}{D} (a - x_1) + \frac{N}{D} \right] w_{x_1}^{\;2} \Big\}. \tag{2.28}$$

The terms $Q[w]$ and $F(\underset{\sim}{x},t)$ in (2.26) are specified as follows:

$$Q[w] = - g\, w_{x_1} . \tag{2.29}$$

This is so, as the follower load g is always tangential to the plate's surface. Consequently, the component (2.29) of the load results which is normal to the undeformed surface of the plate and has no potential. The driving force $F(\underset{\sim}{x},t)$ is distributed over the surface of the plate and is time depending.

From (2.27) follows

$$\frac{d}{dt} \left(\frac{\partial \mathcal{T}}{\partial w_t} \right) = \mu w_{tt} . \tag{2.30}$$

Using (2.20), (2.28) yields

$$\frac{\delta U}{\delta w} = D\Delta^2 w + g[(a - x_1) w_{x_1}]_{x_1} + N\, w_{x_1x_1} , \tag{2.31}$$

where

$$\Delta^2 w = w_{x_1x_1x_1x_1} + 2w_{x_1x_1x_2x_2} + w_{x_2x_2x_2x_2} . \tag{2.32}$$

Using (2.29), (2.30) and (2.31) in (2.26) and rearranging slightly yields the differential equation of the plate

$$\mu\, w_{tt} + D\Delta^2 w + [g(a - x_1) + N]w_{x_1x_1} = F(\underset{\sim}{x},t). \tag{2.33}$$

The terms on the left hand side in (2.33) can be interpreted as follows: the first stems from the inertia, the second term from the elasticity of the plate's material, the third term comprises the contributions of the external loads. The term on the right hand side in (2.33) represents the driving force. Therefore, the equation of motion of an elastic system may be written in a more abstract form as

$$\mu\, w_{tt} + E[w] + S[w] + N[w] = F(\underset{\sim}{x},t). \tag{2.34}$$

In (2.34), E is the "elasticity operator" which differs from system to system. For example,

$$E = \alpha \frac{d^4}{dx^4}$$

for rods having the flexural rigidity α,

$$E = D\Delta^2$$

for plates having the flexural rigidity D, etc.

Operator S is supposed to be selfadjoint, and operator N non-selfadjoint. This holds true if S[w] represents the contributions coming from the conservative components and if N[w] represents the contributions coming from the non-conservative components of the external loads.

Now let it be shown how the boundary conditions can be derived from (2.25) for the chosen example:

Along $x_1 = 0$,

$$\left.\begin{aligned} & dB = -\,dx_2\,, \quad \frac{\partial}{\partial B} = -\frac{\partial}{\partial x_2}\,, \\ & \sin\alpha = 0, \quad \cos\alpha = 1, \\ & \underline{\mathcal{F}}_* = (\Phi_1 - P, 0), \quad \underline{\mathcal{D}}_* = (w, 0). \end{aligned}\right\} \tag{2.35}$$

Along $x_2 = b$,

$$\left.\begin{aligned} & dB = -\,dx\ , \quad \frac{\partial}{\partial B} = -\frac{\partial}{\partial x_1}, \\ & \sin\alpha = 1, \quad \cos\alpha = 0, \\ & \underline{\mathcal{F}}_* = (\Phi_1, 0), \quad \underline{\mathcal{D}}_* = (w, 0). \end{aligned}\right\} \tag{2.36}$$

Along $x_1 = a$,

$$\left.\begin{aligned} & dB = dx_2\,, \quad \frac{\partial}{\partial B} = \frac{\partial}{\partial x_2}\,, \\ & \sin\alpha = 0, \quad \cos\alpha = 1, \\ & \underline{\mathcal{F}}_* = 0, \quad \underline{\mathcal{D}}_* = 0. \end{aligned}\right\} \tag{2.37}$$

Along $x_2 = 0$,

$$\left.\begin{aligned} & dB = dx_1\,, \quad \frac{\partial}{\partial B} = \frac{\partial}{\partial x_1}\,, \\ & \sin\alpha = 1, \quad \cos\alpha = 0, \\ & \underline{\mathcal{F}}_* = \underline{\mathcal{F}}, \quad \underline{\mathcal{D}}_* = \underline{\mathcal{D}}. \end{aligned}\right\} \tag{2.38}$$

Using (2.35) to (2.38) in the boundary integrals in (2.25), the result is

$$- \int\limits_{x_1=0} \left\{ \Phi_2(w)\delta w_n + P[w]\delta w - (\Phi_1(\delta w) - P[\delta w])w - P[w]\delta w \right\} dx_2$$

$$- \int\limits_{x_2=b} \left\{ \Phi_2(w)\delta w_n - \Phi_1(\delta w)w \right\} dx_1 + \int\limits_{x_1=a} \left\{ \Phi_1(w)\delta w + \Phi_2(w)\delta w_n \right.$$

$$\left. - P[w]\delta w \right\} dx_2 + \int\limits_{x_2=0} \left\{ - \Phi_1(\delta w)w - \Phi_2(\delta w)w_n \right\} dx_1 . \tag{2.39}$$

This expression must vanish. Since the variations δw and δw_n are arbitrary, the following conditions result:

$$\Phi_2(w) = 0, \quad w = 0, \quad \text{along } x_1 = 0, \tag{2.40}$$

$$\Phi_2(w) = 0, \quad w = 0, \quad \text{along } x_2 = b, \tag{2.41}$$

$$\Phi_1(w) - P[w] = 0, \quad \Phi_2(w) = 0, \quad \text{along } x_1 = a, \tag{2.42}$$

$$w = 0, \quad w_n = 0, \quad \text{along } x_2 = 0. \tag{2.43}$$

From (2.40) to 2.43), the geometric boundary conditions

$$\left. \begin{array}{l} w = 0, \quad \text{along } x_1 = 0, \quad x_2 = b, \quad x_2 = 0, \\ w_n = 0, \quad \text{along } x_2 = 0. \end{array} \right\} \tag{2.44}$$

can be read off.

Also the dynamic boundary conditions can be derived from (2.40) to (2.42) taking the data for $\sin \alpha$, $\cos \alpha$, and $\partial/\partial B$ into account contained in (2.35) to (2.38) and interpreting Φ_1 and Φ_2 according to (2.22) and (2.23). The result is

$$\frac{\partial \mathscr{U}}{\partial w_{x_1x_1}} = 0, \quad \text{along } x_1 = 0,$$

$$\frac{\partial \mathscr{U}}{\partial w_{x_2x_2}} = 0, \quad \text{along } x_2 = b,$$

$$\left.\begin{aligned} &\frac{\partial \mathcal{U}}{\partial w_{x_1}} - \frac{\partial}{\partial x_1}\left(\frac{\partial \mathcal{U}}{\partial w_{x_1x_1}}\right) - \frac{\partial}{\partial x_2}\left(\frac{\partial \mathcal{U}}{\partial w_{x_1x_2}}\right) - P[w] = 0, \\ &\frac{\partial \mathcal{U}}{\partial w_{x_1x_1}} = 0, \end{aligned}\right\} \quad \text{along } x_1 = a.$$

Using (2.28) and the fact that

$$P[w] = -N\, w_{x_1} \tag{2.45}$$

yields

$$w_{x_1x_1} + \nu\, w_{x_2x_2} = 0, \qquad \text{along } x_1 = 0 \tag{2.46}$$

$$w_{x_2x_2} + \nu\, w_{x_1x_1} = 0, \qquad \text{along } x_2 = b, \tag{2.47}$$

$$\left.\begin{aligned} &w_{x_1x_1x_1} + (2-\nu) w_{x_1x_2x_2} = 0, \\ &w_{x_1x_1} + \nu\, w_{x_2x_2} = 0, \end{aligned}\right\} \quad \text{along } x_1 = a. \tag{2.48}$$

Herewith, it has been shown that principle (2.25) yields indeed the equation of motion and the boundary conditions of the elastic system in question. Henceforth, it shall be assumed that equation of motion and boundary conditions have already been derived in this or in a corresponding way and are, therefore, known.

2.2 Variational equation, the fundamental problem

The state of the elastic system is well defined once the equation of motion , as well as initial and boundary conditions are available. The state is given by the "vector"

$$\underset{\sim}{Q} = (w, w_t, w_{tt}, w_{x_1}, \ldots, w_{x_1x_1}, \ldots) \tag{2.49}$$

or

$$\underset{\sim}{Q} = (q_1, q_2, \ldots q_n), \tag{2.50}$$

respectively, where the q_i, i.e.

$$q_1 = w, \quad q_2 = w_t, \quad \text{etc}, \tag{2.51}$$

are functions of x and t.

Consider a particular state

$$\underline{Q}^\circ = (w^\circ, w_t^\circ, \ldots) = (q_i^\circ) \tag{2.52}$$

whose stability is to be investigated. Let the perturbation $K(\underset{\sim}{x},t)$ be exerted on the system. Then, state $\underline{Q}^\circ$ suffers a variation, the time behaviour of which decides on stability or instability. The stability investigation thus consists in deriving the analytical expression for this variation, the variational equation, and to predict the behaviour of its solution as a function of time.

By virtue of $K(\underset{\sim}{x},t)$, the quantities w° etc, q_i°, respectively, change into

$$w^p = w^\circ + v, \quad w_t^p = w_t^\circ + v_t, \quad w_{tt}^p = w_{tt}^\circ + v_{tt} \quad \text{etc.} \tag{2.53}$$

or

$$q_1^p = q_1^\circ + p_1, \quad q_2^p = q_2^\circ + p_2, \quad q_3^p = q_3^\circ + p_3 \quad \text{etc.} \tag{2.54}$$

respectively. The behaviour of the variations v, v_t, v_{tt} etc, p_i, respectively, decides on the stability of the unperturbed state Q°. The variational equations describing this behaviour are obtained as follows:

Let the original problem be given as

$$\left.\begin{aligned} \Phi(q_i) &= F(\underline{x},t), \\ [\underset{\sim}{\psi}(q_i)]_B &= \underset{\sim}{G}(B,t), \\ [\underset{\sim}{\chi}(q_i)]_{t_o} &= \underset{\sim}{J}(\underset{\sim}{x},t_o), \end{aligned}\right\} \tag{2.55}$$

which are differential equation, boundary conditions, and initial conditions. In (2.55), Φ, $\underset{\sim}{\psi}$, $\underset{\sim}{\chi}$ are known sets of operators.

Now, perturbation $K(\underset{\sim}{x},t)$ may be executed. Then, the first equation in (2.55) must be extended adding the term $K(\underset{\sim}{x},t)$ on the right hand side, and in all equations (2.55), the q_i must be replaced by the q_i^p as given in (2.54). After that operation, the operators Φ, $\underset{\sim}{\psi}$, and $\underset{\sim}{\chi}$ may be expanded according to Taylor. The result is

$$\left.\begin{aligned} &\Phi(q_i^\circ)+\left[\frac{\partial\Phi}{\partial q_i}\right]_{q_i=q_i^\circ} p_i + \dots \frac{1}{n!}\left[\frac{\partial^n\Phi}{\partial q_i^{\,n}}\right]_{q_i=q_i^\circ} p_i^{\,n} = F(\underset{\sim}{x},t)+K(\underset{\sim}{x},t),\\ &\left[\underset{\sim}{\psi}(q_i^\circ)\right]_B+\left\{\left[\frac{\partial\underset{\sim}{\psi}}{\partial q_i}\right]_{q_i=q_i^\circ} p_i + \dots + \frac{1}{n!}\left[\frac{\partial^n\underset{\sim}{\psi}}{\partial q_i^{\,n}}\right]_{q_i=q_i^\circ} p_i^{\,n}\right\}_B = \underset{\sim}{G}(B,t),\\ &\left[\underset{\sim}{\chi}(q_i^\circ)\right]_{t_o}+\left\{\left[\frac{\partial\underset{\sim}{\chi}}{\partial q_i}\right]_{q_i=q_i^\circ} p_i + \dots + \frac{1}{n!}\left[\frac{\partial^n\underset{\sim}{\chi}}{\partial q_i^{\,n}}\right]_{q_i=q_i^\circ} p_i^{\,n}\right\}_{t_o} = J(\underset{\sim}{x},t_o) \end{aligned}\right\} \quad (2.56)$$

However, the q_i° satisfy equations (2.55). Hence,

$$\left[\frac{\partial\Phi}{\partial q_i}\right]_{q_i=q_i^\circ} p_i + \dots + \frac{1}{n!}\left[\frac{\partial^n\Phi}{\partial q_i^{\,n}}\right]_{q_i=q_i^\circ} p_i^{\,n} = K(\underset{\sim}{x},t) \quad (2.57)$$

and

$$\left.\begin{aligned} &\left\{\left[\frac{\partial\underset{\sim}{\psi}}{\partial q_i}\right]_{q_i=q_i^\circ} p_i + \dots + \frac{1}{n!}\left[\frac{\partial^n\underset{\sim}{\psi}}{\partial q_i^{\,n}}\right]_{q_i=q_i^\circ} p_i^{\,n}\right\}_B = 0,\\ &\left\{\left[\frac{\partial\underset{\sim}{\chi}}{\partial q_i}\right]_{q_i=q_i^\circ} p_i + \dots + \frac{1}{n!}\left[\frac{\partial^n\underset{\sim}{\psi}}{\partial q_i^{\,n}}\right]_{q_i=q_i^\circ} p_i^{\,n}\right\}_{to} = 0 \end{aligned}\right\} \quad (2.58)$$

remain. Equation (2.57) is the nonlinear variational equation as the p_i are certain partial differential quotients of w. Equations (2.58) represent homogeneous boundary and initial conditions.

In many cases, one decides to neglect the nonlinear terms in (2.57) and (2.58). Then, one is left with the linear differential equation

$$\left[\frac{\partial\Phi}{\partial q_i}\right]_{q_i=q_i^{\circ}} p_i = \underset{\sim}{K}(x,t) \tag{2.59}$$

and the linear homogeneous boundary and initial conditions

$$\left.\begin{aligned} &\left\{\left[\frac{\partial\underset{\sim}{\psi}}{\partial q_i}\right]_{q_i=q_i^{\circ}} p_i\right\}_B = 0,\\ &\left\{\left[\frac{\partial\underset{\sim}{\chi}}{\partial q_i}\right]_{q_i=q_i^{\circ}} p_i\right\}_{t_o} = 0.\end{aligned}\right\} \tag{2.60}$$

In (2.56) to (2.60), Einstein's summation convention is to be applied to repeated indices.

As an example for the derivation of the variational equation (2.59) consider

$$w^2 + e^{w_x} = \underset{\sim}{F}(x,t),$$

$$w = q_1, \quad v = p_1, \quad w_x = q_2, \quad v_x = p_2.$$

Then,

$$\Phi(q_i) = q_1^2 + e^{q_2}.$$

Hence,

$$\left[\frac{\partial\Phi}{\partial q_1}\right]_{q_i=q_i^{\circ}} = 2q_1^{\circ}, \quad \left[\frac{\partial\Phi}{\partial q_2}\right]_{q_i=q_i^{\circ}} = e^{q_2^{\circ}},$$

and (2.59) reads

$$2q_1^{\circ}p_1 + e^{q_2^{\circ}} p_2 = K(\underset{\sim}{x},t)$$

or

$$2q_1^\circ v + e^{q_2^\circ} v_x = K(\underset{\sim}{x},t).$$

That is indeed a linear differential equation for the variation v as q_1° and $\exp(q_2^\circ)$ are known coefficients.

The conclusion is that when a perturbation $K(\underset{\sim}{x},t)$ changes the unperturbed state $\underset{\sim}{Q}^\circ = (q_i^\circ)$ into the perturbed state $\underset{\sim}{Q}^p = (q_i^\circ + p_i)$, the variational equation

$$\left.\begin{aligned} D[v] &= K(x,t), \\ D[v] &= \left[\frac{\partial\Phi}{\partial q_i}\right]_{q_i=q_i^\circ} p_i, \end{aligned}\right\} \tag{2.61}$$

has to be solved taking corresponding homogeneous initial and boundary conditions into account. It will subsequently be shown, that this general problem can be reduced further to become the so called fundamental problem which then serves as the basis for stability investigations under suitable circumstances.

Let equation (2.61) be solved under the assumption that there exists a Green function $G(\underline{x},\underline{\xi},t,\tau)$. By means of G, the inverse of operator D assumes the form

$$D^{-1} = \int_T \int_V \ldots . \; G(\underline{x},\underline{\xi},t,\tau)d\underline{\xi}\; d\tau. \tag{2.62}$$

Hence, solving (2.61) for v yields

$$v(\underline{x},t) = D^{-1}[K] = \int_T [\int_V K(\underline{\xi},\tau)G(\underline{x},\underline{\xi},t,\tau)d\underline{\xi}]d\tau. \tag{2.63}$$

The first question is how to derive G. The answer is that $K(\underline{x},t)$ has to be chosen as a Dirac-function

$$K(\underline{x},t) = \delta(\underline{x}-\underline{\sigma},\; t-\rho) \tag{2.64}$$

in (2.63). Then, (2.63) yields $v \equiv G$. This is obvious: first, one has

$$v(\underline{x},t) = \int_T [\int_V \delta(\underline{\xi}-\underline{\sigma}, \tau-\rho)G(\underline{x},\underline{\xi},t,\tau)d\underline{\xi}]d\tau. \tag{2.65}$$

Then, using the specific properties of the Dirac-function δ in (2.65), yields subsequently

$$v(\underline{x},t) \equiv G(\underline{x},\underline{\sigma},t,\rho) \equiv G(\underline{x},\underline{\xi},t,\tau), \tag{2.66}$$

i.e. Green's function.

Now, one has a recipe for the derivation of G: Green's function G is obtained as the reaction $v(\underset{\sim}{x},t)$ of the elastic system due to a perturbation $K(\underline{x},t)$ in the form (2.64) of a Dirac-function. In the sense of physics, a Dirac-function is an impulse. Hence, the mathematical formulation of the recipe reads as follows: seek the solution of the problem

$$\left.\begin{array}{l} D(v) = 0, \\ \text{plus homogeneous boundary conditions,} \\ \text{plus the initial conditions: } v(\underline{x},\tau) = 0, \\ \qquad\qquad\qquad\qquad \mu\, v_t(\underline{x},\tau) = \delta(x-\xi). \end{array}\right\} \tag{2.67}$$

The initial conditions in (2.67) are equivalent to the effect of an impulse on the system. Having solved (2.67), the Green function G is obtained.

In the following, it may be assumed that G is known. Then the behaviour of the system under the influence of a perturbation $K(\underline{x},t)$ is given by

$$v(\underline{x},t) = \int_T [\int_V K(\underline{\xi},\tau)G(\underline{x},\underline{\xi},t,\tau)d\underline{\xi}]d\tau. \tag{2.68}$$

The second question is how equation (2.68) may be used for an evaluation of the stability of the system. For this purpose, the following assumptions may be made

1) $K(\underline{\xi},\tau)$ is sign definite as a function of τ,

2) $\int_T K(\underline{\xi},\tau)d\tau = Z(\underline{\xi},T)$, where T is a finite constant and Z a bounded quantity, (A)

3) $G(\underline{x},\underline{\xi},t,\tau)$ is a continuous function of τ.

Assumptions A1) and A2) are for example satisfied if K is a blast for which the sign of the pressure effecting the system does not change and whose duration is limited. Assumption A3) will be satisfied in most of the cases without question.

By virtue of assumptions A1) and A3), the mean value theorem of integration can be applied to (2.68) yielding

$$v(\underline{x},t) = \int_V [G(\underline{x},\underline{\xi},t,T_o]\int_T K(\underline{\xi},\tau)d\tau]d\xi. \tag{2.69}$$

In (2.69), $T_o \in T$ is a mean value of τ and as such a finite quantity. By virtue of assumption A2), equation (2.69) can be rewritten to yield

$$v(\underline{x},t) = \int_V Z(\underline{\xi},T)G(\underline{x},\underline{\xi},t,T_o)d\xi. \tag{2.70}$$

From (2.70) follows that under assumptions A1) to A3), Green's function G above decides on the behaviour of the system with respect to time t and therefore on the stability of the system. Indeed, the transition period will not affect stability under there circumstances:

$$v(\underline{x},t) = \int_V G(\underline{x},\underline{\xi},t,T_o)\int_{\tau=\tau_o}^{t} K(\underline{\xi},\tau)d\tau d\xi \text{ for } \tau_o \leqq t \leqq T,$$

$$v(\underline{x},t) = \int_V G(\underline{x},\underline{\xi},t,T_o)\int_{\tau=\tau_o}^{T} K(\underline{\xi},\tau)d\tau d\xi \text{ for } t \geqq T.$$

But

$$\int_{\tau_o}^{t} K(\xi,\tau)d\tau = Z(\xi,t).$$

Hence,

$$|v(\underline{x},t)| \leqq \int_V |G| Z(\underline{\xi},t) d\underline{\xi} \quad \text{for } \tau_o \leqq t \leqq T,$$

$$|v(\underline{x},t)| \leqq \int_V |G| Z(\underline{\xi},T) d\underline{\xi} \quad \text{for } t \geqq T.$$

Since always $0 \leqq Z(\underline{\xi},t) \leqq Z(\underline{\xi},T)$,

$$|v(\underline{x},t)| \leqq \int_V |G| Z(\underline{\xi},T) d\underline{\xi} \quad \text{at any time.}$$

This inequality shows that the behaviour of v as a function of time can be evaluated without taking the transition period $0 \leqq t \leqq T$ into account.

Having proved that under assumptions (A), the conduct of G is decisive for stability or instability, the next step should be to take a closer look at G itself.

Assume that G is expandable. Then,

$$G(\underline{x},\underline{\xi},t,T_o) = \sum_n G_n(\underline{x},\underline{\xi},t,T_o), \tag{2.71}$$

where

$$G_n = \frac{1}{\mu} u_n(\underline{\xi},T_o) v_n(\underline{x},t). \tag{2.72}$$

By means of (2.71), (2.72) equation (2.70) can be changed into

$$v(\underline{x},t) = \sum_n v_n(x,t) \int_V \frac{u_n(\underline{\xi},T_o) Z(\underline{\xi},T)}{\mu} d\underline{\xi}. \tag{2.73}$$

Introducing the "Fourier coefficient"

$$a_n = \int_V \frac{u_n(\underline{\xi},T_o) Z(\underline{\xi},T)}{\mu} d\underline{\xi}, \tag{2.74}$$

(2.73) can be transformed to read

$$v(\underset{-}{x},t) = \sum_n a_n v_n(\underset{-}{x},t), \tag{2.75}$$

where the $v_n(\underset{-}{x},t)$ are the linearly independent eigensolutions of the fundamental problem

$$\left.\begin{array}{l} D(v) = 0, \\ \text{plus homogeneous boundary conditions.} \end{array}\right\} \tag{2.76}$$

From (2.75) it becomes obvious that the functions v_n decide on the behaviour of the variation v and consequently on the stability of the elastic system. Therefore, let the following assertion be made:

"If the eigensolution v_n of the fundamental problem (2.76) are stable for any n, then the variation v is stable, and so is the elastic system for the unperturbed state whose variation v is being considered."

This assertion is of great importance. If it were correct the stability investigation of the elastic system could be reduced to the following: Solve the fundamental problem (2.76), i.e. seek the eigensolutions of the homogeneous variational equation $D(v) = 0$ for the boundary conditions imposed by the elastic system. Seek the conditions under which these eigensolutions v_n, $n = 1,2,3...$, are stable for any value of n. Such conditions are also the stability conditions of the elastic system itself.

According to (2.70) - (2.75)

$$v = \sum_n a_n v_n$$

is convergent if

$$G = \sum_n G_n$$

is convergent. Assume that the expansion of the Green function G is absolutely and uniformly convergent. Then, (2.75) is also absolutely and uniformly convergent. Furthermore, let any v_n be stable. Then,

$$\left|a_n v_n\right| < \frac{\varepsilon-\eta}{N} \text{ for } t \geqq \tau_o \tag{2.77}$$

Hence,

$$\sum_{n=1}^{N}\left|a_n v_n\right| < \varepsilon - \eta .$$

But because of

$$\left|\sum_{n=1}^{N} a_n v_n\right| \leqq \sum_{n=1}^{N}\left|a_n v_n\right| ,$$

also

$$\left|\sum_{n=1}^{N} a_n v_n\right| < \varepsilon - \eta \text{ for } t \geqq \tau . \tag{2.78}$$

By virtue of the absolute and uniform convergence of (2.75),

$$\sum_{n=N+1}^{\infty}\left|a_n v_n\right| < \eta \text{ for } t \geqq \tau \tag{2.79}$$

and for appropriately chosen N. Then, also

$$\left|\sum_{n=N+1}^{\infty} a_n v_n\right| < \eta \text{ for } t \geqq \tau . \tag{2.80}$$

Combining (2.80) with (2.78) yields

$$\left|\sum_{n=1}^{N} a_n v_n\right| + \left|\sum_{n=N+1}^{\infty} a_n v_n\right| < \varepsilon \text{ for } t \geqq \tau . \tag{2.81}$$

However,

$$\left|\sum_{n=1}^{N} a_n v_n + \sum_{n=N+1}^{\infty} a_n v_n\right| = \left|\sum_{n=1}^{\infty} a_n v_n\right| = \left|v\right| < \left|\sum_{n=1}^{N} a_n v_n\right| + \left|\sum_{n=N+1}^{\infty} a_n v_n\right| . \tag{2.82}$$

Therefore,

$$|v| < \varepsilon \text{ for } t \geqq \tau. \tag{2.83}$$

as can be concluded from (2.81) and (2.82). But that expresses the stability of v. Hence, the following theorem holds true:

"Consider the stability of a certain state of an elastic system. The investigation of stability leads to the variational equation (2.61) for the variation v of the unperturbed state of the elastic system. Assume that the Green function G corresponding to (2.61) and the prescribed boundary conditions is absolutely and uniformly expandable in terms of the eigenfunctions of the <u>fundamental problem</u> (2.76) which involves the homogeneous version of (2.61). Then, the stability of the eigenfunctions v_n of (2.76) for any n implies the stability of the variation v and, thus, the stability of the reference state of the elastic system. It is therefore sufficient for a stability investigation of the elastic system to investigate the stability of the solutions v_n of the fundamental problem (2.76)."

If the elastic system is a conservative one, it is mathematically selfadjoint. Therefore, it can in general be assumed that the Green function involved is symmetric and expandable in an absolute and uniform way. Thus, the previous theorem applies to systems of that kind. If the system in question is however a nonconservative one, involving for example follower forces, then the corresponding mathematical problem is non-selfadjoint, and the Green function concerned is at best convergent in the mean square. In that case the previous theorem does not hold anymore and must be replaced by a weaker one. Information concerning such a situation can be found in [2].

As an example take the elastic beam of length ℓ and linear mass density μ, simply supported at both ends and subjected to the axial compressive force N. Investigate the stability of the lateral deflection $w(x,t)$ which the beam executes under the influence of a driving force $F(x,t)$. Mathematically, these vibrations are described by the partial differential equation

$$\alpha \frac{\partial^4 w}{\partial x^4} + N \frac{\partial^2 w}{\partial x^2} + \mu \frac{\partial^2 w}{\partial t^2} = F(x,t),$$

the boundary conditions

$$w(o,t) = \frac{\partial^2 w(o,t)}{\partial x^2} = w(\ell,t) = \frac{\partial^2 w(\ell,t)}{\partial x^2} = 0,$$

and the initial conditions

$$w(x,\tau) = f(x), \; \mu \frac{\partial w(x,\tau)}{\partial t} = g(x).$$

Let the stability of a certain solution $w°$ be considered. For this purpose substitute everywhere w by $w^p = w° + v$, where v may be caused by the perturbation $K(x,t)$. The result is the set of variational equations

$$\left.\begin{aligned} &\alpha \frac{\partial^4 v}{\partial x^4} + N \frac{\partial^2 v}{\partial x^2} + \mu \frac{\partial^2 v}{\partial t} = K(x,t), \\ &v(o,t) = \frac{\partial^2 v(o,t)}{\partial x^2} = v(\ell,t) = \frac{\partial^2 v(\ell,t)}{\partial x^2} = 0, \\ &v(x,\tau) = 0, \quad \mu \frac{\partial v(x,\tau)}{\partial t} = 0. \end{aligned}\right\} \tag{2.84}$$

For a derivation of Green's function assume that $K(x,t)$ is an impulse. Under that assumption, the previous problem changes into the <u>fundamental problem</u>

$$\left.\begin{aligned} &\alpha \frac{\partial^4 v}{\partial x^4} + N \frac{\partial^2 v}{\partial x^2} + \mu \frac{\partial^2 v}{\partial t} = 0, \\ &v(o,t) = \frac{\partial^2 v(o,t)}{\partial x^2} = v(\ell,t) = \frac{\partial^2 v(\ell,t)}{\partial x^2} = 0. \end{aligned}\right\} \quad (2.85)$$

This is a special case of (2.76). In addition, the initial conditions

$$v(x,\tau) = 0, \quad \mu \frac{\partial v(x,\tau)}{\partial t} = \delta(x - \xi) \quad (2.86)$$

must be satisfied.

The eigensolutions of (2.85) are

$$v_n(x,t) = \frac{1}{\omega_n}\sqrt{\frac{2}{\ell}} \sin \frac{n\pi}{\ell} x \sin \omega_n (t - \tau), \quad (2.87)$$

as can easily be seen.

Hence

$$v(x,t) = \sum_n a_n v_n(x,t),$$

where the a_n must be chosen so that (2.86) is satisfied. Condition $v(x,\tau) = 0$ is immediately fulfilled. The second initial condition requires

$$\frac{\partial v}{\partial t}(x,\tau) = v_t(x,\tau) = \sum_n a_n v_{n,t}(x,t) = \frac{1}{\mu} \delta(x - \xi).$$

This condition is satisfied if the coefficients a_n are chosen such that

$$\int_o^\ell \sum_n a_n v_{n,t}(x,\tau) v_{m,t}(x,\tau) dx = \frac{1}{\mu} \int_o^\ell \delta(x - \xi) v_{m,t}(x,\tau) dx$$

holds true. Since the functions $v_{n,t}$ are orthonormal,

$$a_m = \frac{1}{\mu} v_{m,t}(\xi,\tau)$$

results. Hence,

$$v(x,t) \equiv G(x,\xi,t,\tau) = \sum_n \frac{1}{\mu} v_{n,t}(\xi,\tau) v_n(x,t),$$

which by virtue of (2.87) yields

$$G(x,\xi,t,\tau) = \frac{2}{\mu\ell}\sum_n \frac{\sin\frac{n\pi}{\ell}x \sin\frac{n\pi}{\ell}\xi}{\omega_n} \sin\omega_n(t-\tau). \qquad (2.88)$$

The expansion of G is obviously absolutely and uniformly convergent. Therefore, the solution (2.68) of (2.84) can be replaced with any desired degree of accuracy by

$$v(x,t) = \frac{2}{\mu\ell}\int_T\left[\int_o^\ell K(\xi,\tau)\sum_n \frac{\sin\frac{n\pi}{\ell}x \sin\frac{n\pi}{\ell}\xi}{\omega_n}\sin\omega_n(t-\tau)d\xi\right]d\tau. \qquad (2.89)$$

Let assumptions A1) to A3) be valid for $K(\xi,\tau)$. Then, in place of (2.89)

$$v(x,t) = \frac{2}{\mu\ell}\int_o^\ell Z(\xi,T)\sum_n\left[\sin\frac{n\pi}{\ell}x \sin\frac{n\pi}{\ell}\xi\frac{1}{\omega_n}\sin\omega_n(t-T_o)\right]d\xi. \qquad (2.90)$$

This expression shows that the stability of v depends completely on the eigenvalue ω_n of the fundamental problem (2.85). It is therefore true that the stability investigation could have been reduced to determining the eigensolutions of (2.85) and discussing them.

One can easily show that

$$\omega_n = \omega_{o,n}\left(1 - \frac{N}{N_{E,n}}\right)^{1/2}, \quad \omega_{o,n}^2 = \frac{\alpha}{\mu}\left(\frac{n\pi}{\ell}\right)^4,$$

where $\omega_{o,n}$ is the n-th eigenfrequency of the unloaded beam and $N_{E,n}$ the n-th Euler load (buckling load) of the beam. As long as $N < N_{E,n}$, ω_n is real, and the beam executes stable vibrations as given by (2.90). If $N > N_{E,n}$, ω_n becomes imaginary, for example $\omega_n = i\alpha_n$. Then, (2.90) changes into

$$v(x,t) = \frac{2}{\mu\ell}\int_o^\ell Z(\xi,T)\sum_n\left[\sin\frac{n\pi}{\ell}x \sin\frac{n\pi}{\ell}\xi\frac{1}{\alpha_n}\sin h\,\alpha_n(t-T_o)\right]d\xi,$$

which means that the lateral deflections of the beam have become unstable.

3. Approaches to the Solution

From the previous deliberations it can be concluded that under fairly general assumptions, (expandability of the Green function of the variational equation; sign definiteness, boundedness, finite duration of the perturbation), the stability problem of an elastic system can be reduced to the discussion of eigenvalues and eigensolutions of the fundamental problem. This problem consists of the homogeneous variational equations and appropriate boundary conditions. It is reasonable to restrict oneself to the fundamental problem without great loss of generality. Also, it may for the sake of brevity be assumed, that the fundamental problem has already been derived and is given. Then, the following considerations can be focussed on determining the solution of the fundamental problem. This can be done in various ways which shall be described in the subsequent sections.

3.1 Liapunov's approach

Introduction

This approach consists in using the given fundamental problem directly for an evaluation of stability without integrating it at all. A. M. Liapunov developed his Second Method [6] for the investigation of the stability of discrete dynamical systems. An account of his theory and to a certain degree of its extension has for example been given by J. G. Malkin [7], N. G. Chetayev [8], and W. Hahn [9]. First attempts to apply the theory to continuous systems, i.e., to systems mathematically described in terms of partial differential equations, have been made by V. I. Zubov [10]. Also, A. A. Movchan [11] has contributed essentially to make Liapunov's Second Method workable for continuous systems. A detailed and well understandable report on Movchan's formuation of the Liapunov's theory has been given by R. J Knops and

and E. W. Wilkes [12]. These two authors have in addition published an extensive and thorough monograph on the theory of elastic stability [13] in which they deal among other topics with the reformulation of Liapunov's theory for continuous systems. A concise treatment of Liapunov's Second Method as applied to structural systems can be found in a book by C. L. Dym [14]. This book is last but not least of great value as it contains a large number of references, thus, pointing out for example the works done by R. H. Plaut, P. K. C. Wang, G. A. Hegemier, P. C. Parks, S. M. Holzer, J. A. Walker, etc., in connection with Liapunov's theory and continuous systems. A first step to consider Liapunov's theory together with <u>non-conservative</u> continuous systems was made by P. C. Parks [15]. Further investigations of the stability of nonconservative continuous systems by means of Liapunov's method are due to H. H. E. Leipholz [16]. Although Liapunov himself did not produce that branch of his theory dealing with continuous systems it shall for the sake of brevity always be talked simply of Liapunov's method even when referring to continuous systems.

<u>Functional analytical considerations</u>

Let the concept of "intensified Liapunov stability", as created by H. D. Schräpel [4] be adopted. Then the following can be stated: Consider the "characteristic" w of a system, whose behaviour decides on stability, and assume that w depends on two parameters P_1 and P_2. For stability, require

$$|w(P_1, P_2^*+dP_2) - w(P_1, P_2^*)| < \varepsilon \text{ for all } P_2 \geqq P_1^\circ . \tag{3.1}$$

In (3.1), P_2^* is a prescribed value of P_2, and P_1° is the initial value of P_1. Let P_2 be interpreted as a perturbation which causes the characteristic w of the system to change into w + v by virtue of its increment dP_2, and, therefore,

$$w(P_1, P_2^*+dP_2) = w(P_1, P_2^*) + v(P_1, P_2^*). \tag{3.2}$$

But then, one has

$$|v(P_1, P_2^*)| < \varepsilon \text{ for all } P_1 \geqq P_1^\circ, \tag{3.3}$$

instead of (3.1), in the case of stability.

In many problems concerned with the stability of continuous systems, condition (3.3) proves to be too rigorous for practical applications. Therefore, it may be replaced by another one which is more general in the sense of functional analysis. This may be done by replacing the symbol $|\cdots|$ in (3.3), denoting "absolute value", by a different symbol denoting "norm" in an appropriately defined abstract space. For example, $||\cdots||$ is such a symbol [17]. Then (3.3) changes into

$$||v|| < \varepsilon \text{ for all } P_1 \geqq P_1^\circ. \tag{3.4}$$

For the variation v, the differential equation

$$G[v, v_{x_i}, v_{x_i x_j}, \ldots, v_{P_1}, v_{P_1 P_1}, \ldots, v_{x_i x_j P_1}, \ldots] = 0 \tag{3.5}$$

may have been obtained. Let an abstract space with the elements

$$\mathcal{Q} = \mathcal{Q}(q_1, \ldots, q_n) \tag{3.6}$$

be defined. The coordinates of the "points" $\mathcal{Q}$ are supposed to have the interpretation

$$q_1 = v,\ q_2 = v_{x_i},\ q_3 = v_{x_i x_j},\ \ldots,\ q_n = v_{P_1}. \tag{3.7}$$

If, for example, v were the variation of the deflection of the system and P_1 were the time, then the coordinates q_i of $\mathcal{Q}$ were variations of deflection, slopes, curvatures, and rates of deflection. Hence, each element $\mathcal{Q}$ of the abstract space represents a specific <u>state</u> of the system.

For the unperturbed state, the stability of which is to be investigated, v and all its derivatives are identically equal to zero. Therefore, the image point of this state is the origin of the abstract space. For stability, it is required that the image points $\mathcal{Q}$ of the perturbed states, whose coordinates are the nonvanishing variations and the derivatives of these variations, remain always in the neighbourhood of the origin for all values of P_1. By the chosen norm, a metric is introduced, and in the sense of this metric, the norm gives the distance of any element in the space from the space's origin. Condition (3.4) can therefore be interpreted as follows: If (3.4) is satisfied, points $\mathcal{Q}$ remain in a prescribed, arbitrary small distance from the origin.

A common approach is the following: a <u>functional</u>

$$M[q_i] = C \tag{3.8}$$

is chosen, which is determined by the constant C on the right hand side of (3.8). In addition, two norms must be found which, together with M, satisfy the condition

$$\kappa_{(1)}||\mathcal{Q}||_{(1)} \leqq M \leqq \kappa_{(2)}||\mathcal{Q}||_{(2)} \quad \text{for all } P_1 \geqq P_1{}^{\circ}. \tag{3.9}$$

In (3.9), $\kappa_{(1)}$ and $\kappa_{(2)}$ are appropriately chosen constants. According to the properties of a norm, inequality (3.9) expresses the fact that the functional M is positive definite and admits of an upper bound.

By virtue of (3.9), M can be interpreted as a <u>local metric topology</u> with respect to the origin of the abstract space. Indeed, M measures the distance of the points $\mathcal{Q}$ located on M from the origin. If the distance d of two points $\mathcal{P}$ and $\mathcal{Q}$ is defined as

$$d(\mathcal{P}, \mathcal{Q}) = M(\mathcal{P}) - M(\mathcal{Q}), \tag{3.10}$$

one realizes easily that the triangle inequality [17] p.25,

$$d(\mathcal{O}, \mathcal{P}) \geqq d(\mathcal{O}, \mathcal{Q}) + d(\mathcal{P}, \mathcal{Q}) \tag{3.11}$$

is satisfied as

$$M(\mathcal{P}) \geqq M(\mathcal{Q}) + [M(\mathcal{P}) - M(\mathcal{Q})],$$

if one uses (3.10) in (3.11).

Let now the variation be determined which M suffers following a change of P_1. For this purpose consider

$$\frac{dM}{dP_1} = \frac{\partial M}{\partial q_j} \frac{\partial q_j}{\partial P_1}. \tag{3.12}$$

From (3.5) follows with (3.7) that

$$\frac{\partial q_j}{\partial P_1} = S(q_j), \tag{3.13}$$

where S is a certain expression in the q_j which is prescribed by (3.5). Substituting (3.13) in (3.12) yields

$$\frac{dM}{dP_1} = \frac{\partial M}{\partial q_j} S(q_j). \tag{3.14}$$

Integrating, one is lead to

$$[M]_{P_1} = [M]_{P_1^\circ} + \int_{P_1^\circ}^{P_1} \frac{\partial M}{\partial q_j} S(q_j)\, dP_1. \tag{3.15}$$

Since (3.9) holds true for all $P_1 \geqq P_1^\circ$, the inequality on the right hand side may be considered, and one obtains

$$[M]_{P_1^\circ} \leqq \kappa_{(2)} \|\mathcal{Q}\|_{(2)} < \bar{\varepsilon}, \tag{3.16}$$

where $\bar{\varepsilon}$ is a positive, arbitrarily small constant. If in addition,

$$[M]_{P_1} < \bar{\varepsilon} \tag{3.17}$$

is required to hold for all $P_1 \geqq P_1°$, then M must satisfy according to (3.15) and (3.16) the condition

$$\int_{P_1°}^{P_1} \frac{\partial M}{\partial q_j} S(q_j)\, dP_1 < \bar{\varepsilon}\,(1-\gamma), \quad \gamma < 1 , \tag{3.18}$$

which rules the rate of growth of M. Assuming that (3.18) is satisfied, one can also assume that (3.17) holds true, and (3.9) can be replaced by

$$\kappa_{(1)} ||\mathcal{Q}||_{(1)} \leqq M < \bar{\varepsilon} \text{ for all } P_1 \geqq P_1° . \tag{3.19}$$

From (3.19) follows immediately that

$$||\mathcal{Q}||_{(1)} < \bar{\varepsilon}/\kappa_{(1)} = \varepsilon \text{ for all } P_1 \geqq P_1° , \tag{3.20}$$

where ε is positive and arbitrarily small. But (3.20) indicates that stability prevails, because transforming backward by means of (3.6) and (3.7) into an expression in v and its derivatives, (3.20) changes into

$$||v||_{(1)} < \varepsilon \text{ for all } P_1 \geqq P_1° , \tag{3.21}$$

which corresponds to the stability condition (3.4) in the sense of the specific norm $||\cdots||_{(1)}$.

Summarizing, one can formulate the following theorem of stability:

If a functional M can be found in the abstract q_j - space (space of v and its derivatives) such that the functional is positive definite, bounded according to (3.9), and satisfies the rate of growth condition (3.18), then it follows from (3.16) that (3.19), (3.20) and (3.21) hold true. That means, the unperturbed state w, which suffers the variations v, is stable.

Liapunov's second method

Identifying in the previous section, the parameter P_1 with the time t, assuming that the perturbation P_2 is an impulse which causes a change of the initial values for the motion of the system, and equalizing v with the deviation of the system from an unperturbed state w, one realizes without difficulties that the preceding section on functional analytical considerations includes already all the specific features of Liapunov's theory. One has only to require that condition (3.18) is to be replaced by the specific and corresponding condition

$$\frac{dM}{dP_1} \equiv \frac{dM}{dt} \equiv \frac{\partial M}{\partial q_j} S(q_j) \leqq 0 , \tag{3.22}$$

which is typical for the Liapunov approach.

A more general condition which would make (3.19) to hold true were

$$\frac{dM}{dP_1} \equiv \frac{dM}{dt} = \alpha[\bar{\varepsilon}-M(P_1{}^\circ)]e^{\alpha(P_1{}^\circ-P_1)}, \quad P_1 \equiv t,$$

for any $\alpha > 0$. However, only the classical condition (3.22) shall subsequently be considered.

The procedure suggested by Liapunov's theory is as follows: seek a functional M, the so called Liapunov functional which is positive definite and bounded according to (3.9) and whose derivative with respect to the time t is not positive according to (3.22). If such a functional can be found, one can conclude that stability in the sense of the norm $||\cdots||_{(1)}$ is ensured according to

$$||v||_{(1)} < \varepsilon \text{ for all } t \geqq t_o \tag{3.23}$$

corresponding to (3.21). It may be noted that the nature of the norm in (3.23) depends on the nature of the selected functional M.

All this is essentially different than the situation following from Liapunov's original theory. That theory was valid for spaces with a finite number of coordinates. But in such spaces, all norms are equivalent to the Euclidean norm. It is therefore sufficient for systems with a finite degree of freedoms to choose a Liapunov function M which is positive definite and bounded and whose derivative with respect to the time is not positive for having already the full information on stability. The norm underlying the stability criterion has not to be mentioned specifically, since it can always be assumed to be Euclidean. This is not so for continuous systems. Even if M satisfies the conditions of positive definiteness, boundedness and rate of growth, the induced stability statement is more or less powerful depending on the kind of norm $||\cdots||_{(1)}$ which comes into play through the choice of the functional M.

Conservative systems

The motion of a conservative system can be described in terms of the canonical equations

$$\frac{\delta H}{\delta v} = -\dot{p} \;, \quad \frac{\delta H}{\delta p} = \dot{v} \;, \tag{3.24}$$

and, hence, by means of a first order system. In (3.24),

$$H = \int_V \mathscr{H}(p,\, v,\, v_{x_i},\, \ldots)dV \tag{3.25}$$

is the Hamiltonian, which possesses the density $\mathscr{H}$, V is the volume of the system, v is the variation of the unperturbed state, x_i are spatial coordinates, t is the time and p is the generalized impuls corresponding to v. On the left hand side of (3.24) one encounters so called functional derivatives [18] defined as

$$\frac{\delta H}{\delta v} = \frac{\partial \mathcal{H}}{\partial v} - \frac{\partial}{\partial x_i}\frac{\partial \mathcal{H}}{\partial v_{x_i}} + \frac{1}{2}\left[\frac{\partial^2}{\partial x_i{}^2}\frac{\partial \mathcal{H}}{\partial x_{(i)}x_i} + \frac{\partial^2}{\partial x_j \partial x_k}\frac{\partial \mathcal{H}}{\partial v_{x_j x_k}}\right] \tag{3.26}$$

and

$$\frac{\delta H}{\delta p} \equiv \frac{\partial \mathcal{H}}{\partial p} . \tag{3.27}$$

As an example, consider the rod shown in Figure 6. Its Hamiltonian reads

$$H = \frac{1}{2}\int_0^\ell [\mu\dot{v}^2+\alpha v_{xx}{}^2-g(\ell-x)v_x{}^2]dx , \tag{3.28}$$

where μ is the linear mass density, α the bending stiffness, ℓ the length and v the lateral deflection of the rod. g denotes the uniformly distributed unidirectional compressive load.

Comparing (3.28) with (3.25), and taking

$$p = \mu\dot{v} \tag{3.29}$$

into account, yields

$$\mathcal{H} = \frac{1}{2}\left[\frac{p^2}{\mu} + \alpha v_{xx}^2 - g(\ell-x)v_x^2\right]. \tag{3.30}$$

Applying (3.26) to (3.30) yields

$$\frac{\delta H}{\delta v} = g(\ell-x)v_{xx} - gv_x + \alpha v_{xxxx} , \tag{3.31}$$

and applying (3.27) results in

$$\frac{\delta H}{\delta p} = \frac{p}{\mu} . \tag{3.32}$$

Using (3.31) in (3.24) leads to

$$\dot{p} + g(\ell-x)v_{xx} - gv_x + \alpha v_{xxxx} = 0 . \tag{3.33}$$

With (3.32) in (3.24) one obtains

$$\dot{v} - \frac{p}{\mu} = 0 \, . \tag{3.34}$$

Substituting (3.34) back in (3.33) yields finally

$$\mu\ddot{v} + \alpha v_{xxxx} + g(\ell - x)v_{xx} - gv_x = 0 \, , \tag{3.35}$$

which is the differential equation of the small vibrations of the rod.

The derivative of the Hamiltonian with respect to the time reads

$$\dot{H} = \int_V \left(\frac{\delta H}{\delta v} \dot{v} + \frac{\delta H}{\delta p} \dot{p} \right) dV \, . \tag{3.36}$$

Since the system under consideration is a continuous one, functional derivatives instead of partial derivatives had to be used in (3.36). Applying (3.24) to (3.36) yields

$$\dot{H} \equiv \frac{dH}{dt} \equiv 0 \, . \tag{3.37}$$

Hence, H is a functional satisfying the growth condition (3.22). Therefore, H may be used as a Liapunov functional. A first, immediate conclusion is that the trivial equilibrium position of the rod remains stable as long as H is positive definite.

The total energy of the rod is given by H. Thus, using H as a Liapunov functional yields obviously the connection of Liapunov's stability criterion with the classical energy criterion of stability [19]. Hence, using Liapunov's method in this case appears to be redundant, as, at a first glance, one does not obtain any result which one would not have also obtained by means of the classical criterion. However, this is not quite true. One must also satisfy condition (3.9), which leads one to an additional statement about the kind of norm with respect to which stability is to be expected.

Reference to a norm is missing in the classical theory.

From (3.28) follows

$$H < \frac{1}{2}\int_0^{\ell} (\mu\dot{v}^2 + \alpha v_{xx}^2)dx = \kappa_{(2)}||v||^2_{(2)} , \qquad (3.38)$$

where

$$\kappa_{(2)} = \max\left(\frac{\mu\ell}{2} , \frac{\alpha\ell}{2}\right) \qquad (3.39)$$

and

$$||v||_{(2)} = [\sup|\dot{v}|^2+\sup|v_{xx}|^2]^{1/2} . \qquad (3.40)$$

Also, from (3.28) follows

$$H > \frac{1}{2}\int_0^{\ell} [\alpha v_{xx}^2 - g(\ell-x)v_x{}^2]dx . \qquad (3.41)$$

Let be

$$\int_0^{\ell} (\ell-x)v_x{}^2 dm = m\int_0^{\ell} v_x{}^2 dx , \qquad (3.42)$$

where m is the distance of the centroid of the $v_x{}^2$ - surface from the edge $x = \ell$. It may be possible to guess an analytical approximation of v(x) as one may have an idea of the shape of the deformed rod. Assume for example $v(x) = x^2$. Then,

$$v_x = x$$

and one obtains

$$m = \frac{\int_0^{\ell} (\ell-x)x^2 dx}{\int_0^{\ell} x^2 dx} = \frac{\ell^4/12}{\ell^3/3} = \frac{\ell}{4} .$$

With that value of m, equation (3.42) can be substituted approximately by

$$\int_0^\ell (\ell-x)v_x^{\,2}dx \approx \frac{\ell}{4}\int_0^\ell v_x^{\,2}dx \, . \tag{3.43}$$

Using the inequality of Schwarz [20] yields

$$\int_0^\ell v_{xx}^{\,2}\,dx \geqq \frac{2}{\ell^2}\int_0^\ell v_x^{\,2}dx \, . \tag{3.44}$$

By means of (3.43) and (3.44), inequality (3.41) can be changed into

$$H > \frac{1}{2}\int_0^\ell \left[\alpha - \frac{g\ell^3}{8}\right] v_{xx}^{\,2}\,dx \, . \tag{3.45}$$

For H to remain positive definite, one must require that

$$\alpha - \frac{g\ell^3}{8} > 0 \, ,$$

which leads to

$$g < \frac{8\alpha}{\ell^3} \approx g_{crit} \tag{3.46}$$

Thus, one has obtained a good approximation of the buckling load, whose exact value [21] is

$$g_{crit} = \frac{(1.986)^3\alpha}{\ell^3} = \frac{7.83\alpha}{\ell^3} \, . \tag{3.47}$$

Furthermore, (3.45) can be changed into

$$H > \frac{\bar{\kappa}}{2}\int_0^\ell v_{xx}^{\,2}\,dx \, , \qquad \bar{\kappa} = \alpha - \frac{g\ell^3}{8} > 0 \qquad \text{for } g < g_{crit} \, . \tag{3.48}$$

Applying once more the inequality of Scharz [20], one obtains

$$v^2 < \frac{\ell^3}{2}\int_0^\ell v_{xx}^{\,2}\,dx \, . \tag{3.49}$$

Therefore, (3.48) can be rewritten to yield

$$H > \frac{\bar{\kappa}}{2} \cdot \frac{2}{\ell^3} v^2 \ . \tag{3.50}$$

That is equivalent to

$$H > \kappa_{(1)} ||v||^2_{(1)} \ , \tag{3.51}$$

where $\kappa_{(1)} = \bar{\kappa}/\ell^3$ and

$$||v||_{(1)} = |v| \ . \tag{3.52}$$

By means of (3.38) and (3.51) one obtains

$$\kappa_{(1)} ||v||^2_{(1)} < H < \kappa_{(2)} ||v||^2_{(2)} \quad \text{for } t \geqq t_o \tag{3.53}$$

which corresponds to (3.9). If the motion of the rod is initiated with sufficiently small velocities $\dot{v}$ and curvatures v_{xx}, one can ensure that

$$\kappa_{(2)} ||v||^2_{(2)} = \max\left(\frac{\mu\ell}{2}, \frac{\alpha\ell}{2}\right) [\sup|\dot{v}|^2 + \sup|v_{xx}|^2]^{1/2} < \bar{\varepsilon} \text{ for } t = t_o \ . \tag{3.54}$$

Thus,

$$H < \bar{\varepsilon} \qquad \text{for } t = t_o \tag{3.55}$$

follows from (3.53). But, according to (3.37), $\dot{H} = 0$. Hence, H does not depend explicitely on t and (3.55) must hold true for any time. Consequently,

$$H < \bar{\varepsilon} \qquad \text{for } t \geqq t_o \tag{3.56}$$

Using this result in (3.53) yields

$$\kappa_{(1)} ||v||^2_{(1)} < \bar{\varepsilon} \qquad \text{for } t \geqq t_o \ . \tag{3.57}$$

With $\overline{\varepsilon}/\kappa_{(1)} = \varepsilon^2$ one arrives at

$$||v||^2_{(1)} < \varepsilon^2 \qquad \text{for } t \geqq t_o . \tag{3.58}$$

By virtue of (3.52), (3.58) is equivalent to

$$|v| < \varepsilon \qquad \text{for } t \geqq t_o . \tag{3.59}$$

This statement corresponds to (3.23) and indicates that there is stability.

Summarizing, the following theorem can be formulated: For the rod shown in Figure 6, $\dot{H} = 0$. If, in addition, $g < g_{crit} = 7.82\alpha/\ell^3$, H is positive definite. In that case, the initial condition (3.54) ensures the validity of (3.59) for all time. That means, the lateral deflections of the rod remain small for all time. Hence, the trivial equilibrium position of the rod is stable.

It is noteworthy that not always a norm $||\cdots||_{(1)}$ can be found which ensures stability in the sense of pointwise smallness of the deflections as in the case of the rod. Sometimes one must content oneself with less, e.g., with the smallness of the deflections in the mean square. In such a case, stability does not exclude large deflections locally. Hence, local buckling cannot be ruled out. As an example for the occurrence of such a situation, consider the plate problem treated in the next section.

Nonconservative systems

As an example consider the plate in Figure 7 subjected to nonconservative follower forces which result from the effect that streaming media exert on the plate. In this case, the Hamiltonian cannot be chosen any longer as the Liapunov functional, and the very deficiency of the otherwise powerful method of Liapunov comes to light, namely, the missing of a systematic procedure to arrive at a proper Liapunov function. Like in this example, this functional must be guessed. However, nevertheless, one can be quite successful in finding the functional, as will be seen subsequently.

The variational equation of the plate in Figure 7 carrying non-conservative loads reads

$$\mu\ddot{v} + D\Delta^2 v + q(\ell - x_1)v_{x_1x_1} + \beta\dot{v} = 0 . \tag{3.60}$$

In (3.60), $v(x,t)$ is the variation of the original deflection of the plate, μ its mass density per surface unit, D its flexural rigidity, ℓ its length in the x_1-direction, q the uniformly distributed, tangential follower forces, and β the coefficient of damping.

The variational equation of the corresponding plate subjected to conservative, unidirectional forces as shown in Figure 8 reads

$$\mu\ddot{y} + D\Delta^2 y + q[(\ell - x_1)y_{x_1}]_{x_1} + \beta\dot{y} = 0 . \tag{3.61}$$

Let the relationship

$$\int_S g \frac{\partial h}{\partial x_i} dS = - \int_S h \frac{\partial g}{\partial x_i} dS + \int_E hg \cos(n,x_i) dE \tag{3.62}$$

be introduced, where S denotes the plate's surface, E the edge of the plate, and where g and h are arbitrary functions. By means of (3.62) one can obtain

$$\int_S v(\ell - x_1)v_{x_1x_1} dS = - \int_S v_{x_1}{}^2(\ell - x_1) dS + \int_S vv_{x_1} dS$$

$$+ \int_E v(\ell - x_1)v_{x_1} \cos(n,x_1) dE \tag{3.63}$$

setting in (3.62) $g = (\ell - x_1)v$ and $h = v_{x_1}$.

Assume that the plate is supported along all edges. Then $v \equiv 0$ along E. Therefore, the line integral in (3.63) vanishes.

Applying (3.62) to

$$\int_S v v_{x_1} dS$$

yields

$$\int_S v v_{x_1} dS = \frac{1}{2} \int_E v^2 \cos(n, x_1) dE \ .$$

But, as already mentioned, $v \equiv 0$ along E. Hence,

$$\int_S v v_{x_1} dS = 0 \ ,$$

and (3.63) becomes

$$\int_S v(\ell - x_1) v_{x_1 x_1} dS = - \int_S v_{x_1}{}^2 (\ell - x_1) dS \ . \qquad (3.64)$$

Green's formula reads

$$\int_S u \Delta^2 v dS = \int_S \Delta u \Delta v dS + \int_E \left(u \frac{\partial \Delta v}{\partial n} - \frac{\partial u}{\partial n} \Delta v \right) dE \ . \qquad (3.65)$$

In (3.65) v is the variation of the plate's deflection and u a comparison function, i.e., a function in the same class as v and satisfying the same boundary conditions as v. Then, the line integral on the right hand side of (3.65) vanishes, because along simply supported edges

$$v = u = 0 \ , \qquad \Delta v = \Delta u = 0 \ , \qquad (3.66)$$

and along clamped edges

$$v = u = 0 \ , \qquad \partial v / \partial n = \partial u / \partial n = 0 \ . \qquad (3.67)$$

Therefore, for the plate in Figure 7,

$$\int_S u\Delta^2 v dS = \int_S \Delta u \Delta v dS \ . \tag{3.68}$$

The mean value theorem of integration yields

$$\int_S (\ell - x_1) v_{x_1}{}^2 dS = m \int_S v_{x_1}{}^2 dS \ , \quad 0 < m < \ell \ . \tag{3.69}$$

With (3.64), (3.68) and (3.69), one disposes of three relationships which, together with three subsequent inequalities, form the basis of all considerations to follow.

For the plate in Figure 8 subjected to conservative loads, the static stability criterion (see [20]) and Rayleigh's inequality apply. If q_{crit} is the buckling load of the plate subjected to conservative forces and u is a comparison function, one has

$$q_{crit} \leq R = \frac{D \int_S u\Delta^2 u dS}{-\int_S u[(\ell - x_1)u_{x_1}]_{x_1} dS} \ . \tag{3.70}$$

Using v, i.e., the variation of the deflection of the plate subjected to follower forces, as the comparison function in (3.70) yields

$$q_{crit} \leq \frac{D \int_S v\Delta^2 v dS}{-\int_S v[(\ell - x_1)v_{x_1}]_{x_1} dS} \ . \tag{3.71}$$

However, by virtue of (3.62)

$$-\int_S v[(\ell - x_1)v_{x_1}]_{x_1} dS = \int_S (\ell - x_1) v_{x_1}{}^2 dS + \int_E v(\ell - x_1) v_{x_1} \cos(n, x_1) dE \ ,$$

which changes into

$$-\int_S v[(\ell - x_1)v_{x_1}]_{x_1} dS = \int_S (\ell - x_1) v_{x_1}{}^2 dS \ , \tag{3.72}$$

since $v \equiv 0$ along E. Moreover, from (3.68), setting there $u \equiv v$, follows

$$\int_S v\Delta^2 v dS = \int_S (\Delta v)^2 dS \ . \tag{3.73}$$

Using (3.72) and (3.73) in (3.71) yields the first of the basic inequalities, that is to say

$$D \int_S (\Delta w)^2 dS \geqq q_{crit} \int_S (\ell - x_1) v_{x_1}{}^2 dS \ . \tag{3.74}$$

A second inequality follows from Poincares' inequality

$$\int_S u^2 dS \leqq A \int_S (u_{x_1}{}^2 + u_{x_2}{}^2) dS + B \left[\int_S u\, dS\right]^2 \ .$$

Setting consecutively $u \equiv v_{x_1}$ and $u \equiv v_{x_2}$ yields two inequalities which may be combined to yield

$$\int_S (v_{x_1}{}^2 + v_{x_2}{}^2) dS \leqq A \int_S (v_{x_1x_1}{}^2 + 2v_{x_1x_2}{}^2 + v_{x_2x_2}{}^2) dS \ . \tag{3.75}$$

In (3.75), A is an appropriately chosen constant. Inequality (3.75) holds true, because due to the support of the plate along all edges,

$$\int_S v_{x_i} dS = \int_E v \cos(n, x_i) dE = 0 \ , \qquad i = 1,2.$$

A third inequality follows from a theorem by Sobolev (see [12], Section 4.3.3, Theorem 1, p. 313). According to it,

$$\sup|v|^2 \leqq m_1{}^2 \widetilde{H}^2 \ ,$$

where m_1 is an appropriately chosen constant and $\widetilde{H}^2$ is the expression

$$\widetilde{H}^2 = \int_E v^2 dE + \int_S (v_{x_1x_1}{}^2 + 2v_{x_1x_2}{}^2 + v_{x_2x_2}{}^2) dS \ .$$

Since for the plate under consideration $v \equiv 0$ along E, it follows that

$$\sup|v|^2 \leqq m_1{}^2 \int_S (v_{x_1x_1}{}^2 + 2v_{x_1x_2}{}^2 + v_{x_2x_2}{}^2) dS \ . \tag{3.76}$$

As a Liapunov function let

$$M = H + \int_S \left(\frac{\beta^2}{4\mu} v^2 + \frac{\beta}{2} v\dot{v} \right) dS \tag{3.77}$$

be introduced. In (3.77), H is the Hamiltonian of the plate, that is to say

$$H = \int_S \left\{ \frac{\mu}{2} \dot{v}^2 + \frac{D}{2} \left[(\Delta v)^2 - \frac{q}{D} (\ell - x_1) v_{x_1}{}^2 \right] \right\} dS \ . \tag{3.78}$$

Differentiating (3.77) with respect to the time yields

$$\dot{M} = \dot{H} + \int_S \left[\frac{\beta^2}{2\mu} v\dot{v} + \frac{\beta}{2} \dot{v}^2 + \frac{\beta}{2} v\ddot{v} \right] dS \ . \tag{3.79}$$

The equations of Hamilton of the plate subjected to follower forces read

$$\frac{\delta H}{\delta w} = -p - qv_{x_1} - \beta\dot{v} \ ,$$

$$\frac{\delta H}{\delta p} = \dot{v} \ .$$

Hence,

$$\dot{H} = \int_S \left(\frac{\delta H}{\delta w} \dot{v} + \frac{\delta H}{\delta p} \dot{p} \right) dS = -q \int_S v_{x_1} \dot{v} dS - \beta \int_S \dot{v}^2 dS \ . \tag{3.80}$$

Moreover, using (3.60), one obtains

$$\frac{\beta}{2} \int_S v\ddot{v} dS = - \frac{\beta D}{2\mu} \int_S v\Delta^2 v dS - \frac{\beta q}{2\mu} \int_S v(\ell - x_1) v_{x_1 x_1} dS - \frac{\beta^2}{2\mu} \int_S v\dot{v} dS \ . \tag{3.81}$$

This relationship can be transformed into

$$\frac{\beta}{2} \int_S v\ddot{v} dS = -\int_S \left[\frac{\beta D}{2\mu} (\Delta v)^2 - \frac{\beta q}{2\mu} (\ell - x_1) v_{x_1}{}^2 + \frac{\beta^2}{2\mu} v\dot{v} \right] dS \tag{3.82}$$

by means of (3.73) and (3.64).

Using (3.80) and (3.82) in (3.79) yields

$$\dot{M} = -\int_S \left[\frac{\beta}{2} \dot{v}^2 + \frac{\beta D}{2\mu} (\Delta v)^2 - \frac{q\beta}{2\mu} (\ell - x_1) v_{x_1}{}^2 + q v_{x_1} \dot{v} \right] dS. \tag{3.83}$$

For the investigation of stability, let the norms

$$||v||_{(1)} = \sup|v| \tag{3.84}$$

and

$$||v||_{(2)} = [\sup|\dot{v}|^2 + \sup|\Delta v|^2 + \sup|v|^2]^{1/2} \tag{3.85}$$

be introduced.

The Liapunov functional M is positive definite if

$$\int_S \left[\frac{\mu}{2} \dot{v}^2 + \frac{\beta^2}{4\mu} v^2 + \frac{\beta}{2} v\dot{v} \right] dS > 0 \tag{3.86}$$

and if

$$\int_S \left[(\Delta v)^2 - \frac{q}{D} (\ell - x_1) v_{x_1}{}^2 \right] dS > 0 \, . \tag{3.87}$$

Condition (3.86) is always satisfied because the integrand is a positive definite form. This follows from the fact that its discriminant is $\beta^2/16$, i.e., it is positive.

Condition (3.87) can be rewritten by virtue of (3.74) to yield

$$\int_S \left[(\Delta v)^2 - \frac{q}{D} (\ell - x_1) v_{x_1}{}^2 \right] ds > \frac{q_{crit} - q}{D} \int_S (\ell - x_1) v_{x_1}{}^2 dS > 0 \, . \tag{3.88}$$

This inequality holds true if

$$q < q_{crit} \, . \tag{3.89}$$

The conclusion is that M is positive definite if the stability condition (3.89) is satisfied.

Taking (3.86) into account, one concludes from (3.77) and (3.78) that

$$M > \frac{D}{2} \int_S \left[(\Delta v)^2 - \frac{q}{D} (\ell - x_1) v_{x_1}{}^2 \right] dS \; . \qquad (3.90)$$

Considering the support conditions of the plate,

$$\int_S D(1-\nu)(v_{x_1x_2}{}^2 - v_{x_1x_1} v_{x_2x_2}) dS = 0 \; ,$$

where ν is Poisson's ratio. Therefore, (3.90) can be changed into

$$M > \frac{D}{2} \int_S \left[(\Delta v)^2 + 2(1-\nu)(v_{x_1x_2}{}^2 - v_{x_1x_1} v_{x_2x_2}) - \frac{q}{D} (\ell - x_1) v_{x_1}{}^2 \right] dS \; . \qquad (3.91)$$

Since $\nu < 1$,

$$\int_S [(\Delta v)^2 + 2(1-\nu)(v_{x_1x_2}{}^2 - v_{x_1x_1} v_{x_2x_2})] dS > \int_S [(1-\nu)(v_{x_1x_1}{}^2 + 2v_{x_1x_2}{}^2 + v_{x_2x_2}{}^2] dS \; . \qquad (3.92)$$

Using (3.92) in (3.91), using in addition (3.69) and subtracting on the right hand side the positive term

$$\int_S \frac{q}{D} m \, v_{x_2}{}^2 dS \; ,$$

the new inequality

$$M > \frac{D}{2} \int_S [(1-\nu)(v_{x_1x_1}{}^2 + 2v_{x_1x_2}{}^2 + v_{x_2x_2}{}^2) - q \frac{m}{D} (v_{x_1}{}^2 + v_{x_2}{}^2)] dS$$

follows from (3.91). By means of (3.75) it may be transformed into

$$M > \frac{D}{2} \left[(1-\nu) - \frac{qmA}{D} \right] \int_S (v_{x_1x_1}{}^2 + 2v_{x_1x_2}{}^2 + v_{x_2x_2}{}^2) dS \; .$$

Finally, one obtains by virtue of (3.76) and (3.84)

$$M > \kappa_{(1)} ||v||^2_{(1)} \; , \qquad (3.93)$$

where

$$\kappa_{(1)} = \frac{D}{2m_1^2}\left[(1-\nu) - \frac{qmA}{D}\right]. \tag{3.94}$$

Since $\kappa_{(1)}$ must be positive, one has to admit of the second stability condition

$$q < \frac{D(1-\nu)}{mA} = q^* \tag{3.95}$$

besides (3.89).

If one wishes to avoid condition (3.95), one has to use

$$||v||_{(1,1)} = [\int_S v^2 dS]^{1/2} \tag{3.96}$$

as a norm instead of (3.84). However, the stability criterion based on (3.96) is weaker than the one based on (3.84): $||v||_{(1)}$ ensures <u>pointwise</u> smallness of the plate's deflections, while $||v||_{(1,1)}$ ensures only smallness of the deflections <u>in the mean square</u>. Hence, local buckling of the plate is not excluded by (3.96). In order to show how to proceed when using $||v||_{(1,1)}$, refer back to (3.90). Applying (3.74) to (3.90) yields

$$M > \frac{1}{2}(q_{crit}-q) \int_S (\ell-x_1) v_{x_1}{}^2 dS\ . \tag{3.97}$$

Now, the inequality of Friedrichs,

$$\int_S u^2 dS \leqq \ell \int_S u_{x_1}{}^2 dS\ ,$$

may be used. It holds true for all functions vanishing along the contour of the plate. But this is the case for v due to the support of the plate along all edges. Therefore, one may set $u \equiv v$ in order to obtain

$$\int_S v^2 dS \leqq \ell \int_S v_{x_1}{}^2 dS\ . \tag{3.98}$$

Using (3.69) together with (3.98) in (3.97) yields

$$M > \frac{m}{2\ell}(q_{crit}-q) \int_S v^2 dS\ . \tag{3.99}$$

This is equivalent to

$$M > \kappa_{(1,1)} ||v||^2_{(1,1)} , \tag{3.100}$$

where

$$\kappa_{(1,1)} = m(q_{crit} - q)/2\ell \tag{3.101}$$

is positive, if the stability condition (3.89) is satisfied.

Now, let an upper bound for M be derived. Because of

$$\frac{\mu}{2} \dot{v}^2 + \frac{\beta}{2} v\dot{v} + \frac{\beta^2}{4\mu} v^2 < \mu\dot{v}^2 + \frac{\beta^2}{2\mu} v^2 ,$$

and suppressing the negative term in (3.78), (3.77) yields

$$M < \int_S \left[\mu\dot{v}^2 + \frac{\beta^2}{2\mu} v^2 + \frac{D}{2} (\Delta v)^2 \right] dS . \tag{3.102}$$

Defining

$$\kappa_{(2)} = \max \left[S\mu, \ \frac{\beta^2 S}{2\mu} , \ \frac{DS}{2} \right] \tag{3.103}$$

and taking (3.85) into account,

$$M < \kappa_{(2)} ||v||^2_{(2)} \tag{3.104}$$

follows from (3.102).

Finally, $\dot{M}$ shall be proved to be negative definite, which implies that M never increases.

Applying (3.74) to (3.73) yields

$$\dot{M} < - \int_S \left[\frac{\beta}{2} \dot{v}^2 + \frac{\beta}{2\mu} (q_{crit} - q)(\ell - x_1) v_{x_1}{}^2 + q v_{x_1} \dot{v} \right] dS . \tag{3.105}$$

Using (3.69), (3.105) changes into

$$\dot{M} < - \int_S \left[\frac{\beta}{2} \dot{v}^2 + \frac{\beta m}{2\mu} (q_{crit} - q) v_{x_1}{}^2 + q v_{x_1} \dot{v} \right] dS . \tag{3.106}$$

$\dot{M}$ is negative definite, if the integrand in (3.106) is positive definite.

That is the case if the discriminant

$$\frac{\beta}{2}\left[\frac{\beta m}{2\mu}(q_{crit}-q)\right]-\frac{q^2}{4}$$

of the integrand is positive. For that to be true, one must require that

$$\frac{\beta^2 m}{4\mu}(q_{crit}-q)>\frac{q^2}{4},$$

which leads to the stability condition

$$\beta > q\left[\frac{\mu}{m(q_{crit}-q)}\right]^{1/2} . \tag{3.107}$$

(3.107) is a condition for the damping coefficient β.

The stability conditions (3.89), (3.95), and (3.107) are displayed in Figures 9 and 10. In the case of Figure 9, stability is interpreted in the sense of norm $||v||_{(1)}$, i.e., in the sense of <u>pointwise</u> smallness of the deflections. In the case of Figure 10, stability is understood in the sense of norm $||v||_{(1,1)}$, i.e., in the sense of smallness of the deflections in the <u>mean square</u>.

Now, the stability of the plate can be discussed: first, it may be assumed that (3.89), (3.95), and (3.107) are all satisfied. Then, one is allowed to work with the norms $||v||_{(1)}$ and $||v||_{(2)}$. At the time $t = t_o$, the varied motion of the plate may begin with sufficiently small values of the velocity $\dot{v}$, the curvature Δv and the deflection v, i.e.,

$$M < \kappa_{(2)}||v||^2_{(2)} < \bar{\varepsilon} \qquad \text{for } t = t_o . \tag{3.108}$$

But $\dot{M}$ is negative definite, and M therefore does not increase. Hence,

$$M < \kappa_{(2)}||v||^2_{(2)} < \bar{\varepsilon} \qquad \text{for } t \geqq t_o . \tag{3.109}$$

Moreover, (3.93) may be used. Combining that inequality with (3.109) yields

$$\kappa_{(1)}||v||^2_{(1)} < \kappa_{(2)}||v||^2_{(2)} < \bar{\varepsilon} \qquad \text{for } t \geqq t_o . \tag{3.110}$$

But then,

$$||v||_{(1)} < \sqrt{\frac{\bar{\varepsilon}}{\kappa_{(1)}}} = \varepsilon \qquad \text{for } t \geqq t_o . \tag{3.111}$$

Because of (3.84) this is equivalent to

$$\sup|v| < \varepsilon \qquad \text{for } t \geqq t_o , \tag{3.112}$$

which is stability. The conclusion is that when satisfying the conditions (3.89), (3.95), and (3.107) for the load and the coefficient of damping, and if observing the initial condition (3.108), the motion of the plate is stable, because according to (3.112), the deflections remain sufficiently small at all time.

Assuming that (3.89) and (3.107) and not (3.95) are satisfied, one must use the norms $||v||_{(1,1)}$ and $||v||_{(2)}$. Everything else runs correspondingly. Again, one has the initial condition (3.108), and because of M being negative definite, also (3.109) remains true. In the next step one must however use (3.100) instead of (3.93). Then,

$$\kappa_{(1,1)}||v||^2_{(1,1)} < \kappa_{(2)}||v||^2_{(2)} < \bar{\varepsilon} \qquad \text{for } t \geqq t_o \tag{3.113}$$

and

$$||v||^2_{(1,1)} < \frac{\bar{\varepsilon}}{\kappa_{(1,1)}} = \varepsilon \qquad \text{for } t \geqq t_o . \tag{3.114}$$

Because of (3.96), inequality (3.114) is equivalent to

$$\int_S v^2 dS < \varepsilon \qquad \text{for } t \geqq t_o . \tag{3.115}$$

This is a weaker stability criterion as the previous one. It states only that stability prevails in the sense of the smallness of the deflections v in the mean square, if (3.89) and (3.107) but not (3.95) are satisfied.

As a conclusion of this section, it may be mentioned that condition (3.94) admits of the following interpretation: If the extension of the plate in the x_2-direction is smaller than in the x_1-direction, one may set $A = \ell^2$. Moreover, $m < \ell$. Let $1-\nu \approx 2/3$. Then (3.95) yields

$$q^* \approx \frac{2D}{3\ell^3},$$

which will be smaller than q_{crit} in general. An estimate of this kind has been assumed to hold in Figure 9.

More examples, including pseudo-conservative and nonlinear systems treated by means of Liapunov's Second Method, can be found in [2] and [22].

3.2 Energy approach

In this case, intermediate integrals of the fundamental problem, like the energy integral, are used for the stability investigation.

In [23], H. Ziegler gave a critical evaluation of the range of applicability of the various classical approaches to the stability problem of elastic systems. Since then it has become a more or less explicit credo that the energy method does not apply to systems which are nonconservative, e.g., subjected to follower forces. This is of course not true, literally. It is correct that the classical variational methods based on a consideration of the Hamiltonian or the potential energy only are not applicable to non-conservative systems. However, any system, also a nonconservative one, admits of an energy formulation. Therefore, an energy method in the broadest sense is indeed applicable to systems of this kind. That shall be shown in the following, where for the sake of simplicity it will be assumed that the original continuous elastic system to be investigated has been discretized to become a lumped mass system.

Let q be the variation of the deflection and p be the corresponding generalized momentum of the system. The nonconservative force acting on the system shall be denoted by N. Hamilton's equations read

$$\partial H/\partial q - N = -\dot{p}, \qquad \partial H/\partial p = \dot{q}. \tag{3.116}$$

By virtue of (3.116)

$$(\partial H/\partial q - N)\dot{q} + (\partial H/\partial p)\dot{p} = 0 \tag{3.117}$$

is an identity which leads to

$$dH/dt - N\dot{q} = 0 . \tag{3.118}$$

Integrating (3.118) with respect to time t yields the energy equation

$$H - \int N\dot{q}\, dt = E = \text{const.} \tag{3.119}$$

Relationship (3.119) can be used to determine the phase curves and, in connection with these, the so called <u>heteroclines</u>. For this purpose consider

$$dE = dH - Ndq = 0 \text{ or } (\partial H/\partial q)dq + \left(\frac{\partial H}{\partial p}\right) dp - Ndq = 0 . \tag{3.120}$$

Observing that there is a <u>destabilizing parameter</u> C (e.g., a load factor) involved, (3.120) can be rewritten to yield

$$\left(\frac{\partial H}{\partial q} - N\right) + \frac{\partial H}{\partial p}\frac{dp}{dq} = \phi(q, p, C) = 0 . \tag{3.121}$$

This is the heterocline, whose behaviour under the influence of C indicates stability or instability.

The identity

$$d\phi = \frac{\partial \phi}{\partial q} dq + \frac{\partial \phi}{\partial \omega} d\omega + \frac{\partial \phi}{\partial C} dC = 0, \tag{3.122}$$

which follows from (3.121), shall be rewritten under the assumptions

$$\left.\begin{aligned} &H = H(q, p, C),\ \partial H/\partial p = \dot{q} = g(\omega, q), \\ &\partial p/\partial q = f(\omega, q), \\ &\frac{\partial H}{\partial p}\frac{dp}{dq} = gf = G(\omega, q),\ N = N(\omega, q, C). \end{aligned}\right\} \quad (3.123)$$

In (3.123), ω is the frequency of the perturbation inflicted motion of the system, i.e., it is assumed that

$$q = Ae^{\omega t},\quad \dot{q} = \omega Ae^{\omega t} = \omega q. \quad (3.124)$$

Using (3.123) in (3.121) and (3.122) yields

$$\left(\frac{\partial^2 H}{\partial q^2} + \frac{\partial G}{\partial q} - \frac{\partial N}{\partial q}\right) dq + \left(\frac{\partial G}{\partial \omega} - \frac{\partial N}{\partial \omega}\right) d\omega + \left(\frac{\partial^2 H}{\partial q \partial C} - \frac{\partial N}{\partial C}\right) dC = 0\ . \quad (3.125)$$

As a stability criterion, the "deformation velocity" $\mathscr{V} = dq/dC$ in the phase space becoming <u>indefinite</u> at inserting instability, shall be used. As can be read off (3.125), $\mathscr{V} = \infty$ is only possible for

$$\frac{\partial^2 H}{\partial q^2} + \frac{\partial G}{\partial q} - \frac{\partial N}{\partial q} = 0. \quad (3.126)$$

(3.126) is a <u>load-frequency relationship</u> which graphically yields the <u>eigenvalue curve</u>.

But, also $d\omega/dC = \infty$ indicates instability. Hence, again by virtue of (3.125), one can conclude that this happens for

$$\partial G/\partial \omega - \partial N/\partial \omega = 0\ . \quad (3.127)$$

Also,

$$\omega = 0 \quad (3.128)$$

is a possibility at inset of instability, if there are <u>divergent</u> buckling loads. Conditions (3.127) or (3.128) are the <u>stability conditions</u>. (3.126) combined with (3.127) or (3.128) yields the <u>critical value</u> of C.

Finally, taking into account that $d\omega \equiv 0$ for fixed values of ω, (3.125) yields

$$dC/dq = -[\partial^2H/\partial q^2 + \partial G/\partial q - \partial N/\partial q].[\partial^2H/\partial q\partial C - \partial N/\partial C]^{-1} \qquad (3.129)$$

which may be integrated in order to yield the load-deflection curve for a post-buckling behaviour investigation.

It is obvious, that the classical, statical case is included in the preceding theory. In that case, $\omega \equiv 0$, $p \equiv 0$, $\dot{q} \equiv 0$, $N \equiv 0$. Thus, $\partial H/\partial q \equiv \partial U/\partial q$, $\partial H/\partial p \equiv 0$, $dp/dq \equiv 0$, $G \equiv 0$, where U is the potential energy. Then, equations (3.121), (3.125), (3.126) and (3.129) change, respectively, into, $\partial U/\partial q = 0$, which is the equilibrium condition, $(\partial^2U/\partial q^2)dq + (\partial^2U/\partial q\partial C) = 0$, $\partial^2U/\partial q^2 = 0$, which expresses the fact that instability occurs at a critical point of the U-surface, $\partial C/\partial q = -(\partial^2U/\partial q^2)(\partial^2U/\partial q\partial C)^{-1}$, which integrated, yields the load deflection curve.

As an example consider the system shown in Figure 11. At hinge A, a restoring moment, and viscous damping is acting. One has to use

$$H = \frac{1}{2}(p^2)/ma^2 + \frac{1}{2}cq^2 - P\ell(1 - \cos q),$$

$$\partial H/\partial p = p/ma^2 = \dot{q} = \omega q = -g(\omega, q),$$

$$dp/dq = d(ma^2\omega q)/dq = ma^2\omega = f(\omega, q),$$

$$G(\omega, q) = g.f = ma^2\omega^2 q,$$

$$N = -\beta\dot{q} = -\beta\omega q, \quad C \equiv P.$$

Hence, (3.126) yields the load frequency relationship,

$$c - P\ell\cos q + ma^2\omega^2 + \beta\omega = 0. \qquad (3.130)$$

From (3.127) follows

$$2ma^2\omega q + \beta q = 0. \qquad (3.131)$$

In addition, $\omega = 0$ may also be taken into account besides (3.131). One may combine (3.131) or the condition $\omega = 0$ with (3.130) in order to find $C_{crit} \equiv P_{crit}$. In this example it turns out to be necessary to use $\omega = 0$ together with (3.130). One thus arrives at

$$P_{crit} = c/\ell \cos q, \quad q \equiv 0. \tag{3.132}$$

Moreover, from (3.129) follows

$$dP/dq = (c - P\ell \cos q)(\ell \sin q)^{-1} + (ma^2\omega^2 + \beta\omega)(\ell \sin q)^{-1}. \tag{3.133}$$

This relationship may be integrated in the neighbourhood of P_{crit}, i.e., at $\omega = 0$, to yield

$$P = cq/\ell \sin q \tag{3.134}$$

for the post-buckling investigation of the system.

A more detailed report including many examples, also the resonance phenomenon of nonlinear vibrations, can be found in [2] and [24]. In those publications, reference has also been made to systems with multiple degrees of freedom.

3.3 Modal approach, discretization, algebraization

As pointed out and justified earlier, it is in many cases sufficient to decide on stability of an elastic system by studying the eigensolution and eigenfunctions of the fundamental problem (2.76). Beside of applying Liapunov's method directly to it or to use the corresponding equations of Hamilton and the intermediate energy integral, one may turn to a third approach which is very common and convenient. It consists in developing the solution of the fundamental problem into a series in terms of a complete set of coordinate functions satisfying the boundary conditions of the fundamental problem.

The method shall be discussed under the assumption that a separation of variables is possible in the fundamental problem. If this is not the case, integral transformations and operations must be applied on which has been reported in [2].

The starting point is the fundamental problem

$$\left.\begin{aligned} &D[v(\underset{\sim}{x},t)] = 0,\\ &\{\underset{\sim}{U}[v(\underset{\sim}{x},t)]\}_B = 0. \end{aligned}\right\} \tag{3.135}$$

The first line represents the differential equation, the second line the boundary conditions. $\underset{\sim}{U}$ is a "vector" of operators acting on $\underset{\sim}{x}$ only. The subscript B indicates that the expressions in the second line are to be taken at the boundary B of the elastic system. The differential operator D is assumed to be of the form

$$D(v) = L_{\underset{\sim}{x}}(v) + M_t(v). \tag{3.136}$$

The operator L acts on $\underset{\sim}{x}$ alone, and the operator M acts on t. Hence, (3.135) can be rewritten to yield

$$\left.\begin{aligned} &L_{\underset{\sim}{x}}(v) + M_t(v) = 0,\\ &\{\underset{\sim}{U}_{x}(v)\}_B = 0. \end{aligned}\right\} \tag{3.137}$$

This problem admits a <u>product solution</u>

$$v(\underset{\sim}{x},t) = \phi(\underset{\sim}{x})\psi(t) \tag{3.138}$$

in which the variables are separated.

Using (3.138) in (3.137) yields

$$\left.\begin{aligned} &L_{\underset{\sim}{x}}(\phi)\psi + M_t(\psi)\phi = 0,\\ &\{\underset{\sim}{U}_{\underset{\sim}{x}}(\phi)\}_B\ \psi = 0. \end{aligned}\right\} \tag{3.139}$$

Since neither ϕ nor ψ can vanish identically if the solution (3.138) does in fact exist,

$$\frac{L_{\underline{x}}(\phi)}{\phi} = - \frac{M_t(\psi)}{\psi} \tag{3.140}$$

and

$$\{\underset{\sim}{U}_{\underline{x}}(\phi)\}_B = 0 \tag{3.141}$$

follow from (3.139).

Relationship (3.140) can only hold true if

$$\frac{L_{\underline{x}}(\phi)}{\phi} = - \frac{M_t(\psi)}{\psi} = \omega, \tag{3.142}$$

where ω is a quantity that depends neither on $\underline{x}$ nor on t. It is the <u>eigenvalue</u> of the problem. Other, different eigenvalues may be implicitely contained in L and M. To each eigenvalue ω there is an <u>eigensolution</u> ϕ.

The procedure to be followed is this: First, the boundary-eigenvalue problem

$$\left.\begin{aligned} L_{\underline{x}}(\phi) - \omega\phi &= 0, \\ \{\underset{\sim}{U}_{\underline{x}}(\phi)\}_B &= 0 \end{aligned}\right\} \tag{3.143}$$

must be solved yielding the eigenvalues and eigensolutions

$$\omega_i, \ \phi_i(\underline{x}), \quad i = 1,2,3,... \tag{3.144}$$

By means of these eigenvalues, the differential equations

$$M_t(\psi) + \omega_i \psi = 0, \quad i = 1,2,3,..., \tag{3.145}$$

can be solved. Thus, the functions

$$\psi_i(\underline{x}), \quad i = 1,2,3,..., \tag{3.146}$$

are determined. By means of (3.144), (3.146), and (3.138), the particular solution

$$v_n(\underline{x},t) = \phi_n(\underline{x})\psi_n(t) \tag{3.147}$$

of the fundamental problem (3.135) can be constructed which are to be used for the evaluation of stability. This is to be done in the following way: Actually,

$$v_n(\underline{x},t,\lambda,\omega) = \phi_n(x,\lambda,\omega)\psi(t,\omega), \tag{3.148}$$

where ω is the already mentioned eigenvalue, and λ is a structural parameter characteristic for the elastic system under investigation. As follows from (3.148) the stability of the elastic system is decided upon by that of $\psi(t,\omega)$. The stability of ψ depends on the behaviour of ω. For certain ranges of ω values, stability is lost. These ranges are determined by discussing the

$$\omega = \omega(\lambda) \tag{3.149}$$

relationship which may be plotted to yield the eigenvalue curve of the problem. The behaviour of ω is controlled by the structural parameter λ. As already mentioned, it is the real aim of the stability investigation to determine the critical values λ_{crit} of the control parameter λ for which ω behaves such that stability is lost.

A solution of (3.143) may not be obtained easily. For this reason one may turn to the method of Galerkin as an approximate method. In this case, ϕ itself is supposed to exist in form of a series. This assumption, which leads subsequently to working with a finite number of coordinates although the elastic system is a continuous one, can be viewed as a discretization of the continuous system.

Corresponding to (3.143), the auxiliary problem

$$\left.\begin{aligned} &S_{\underline{x}}(f_i) = \Omega_i f_i, \\ &\{\underset{\sim}{U}_{\underline{x}}(f_i)\}_B = 0 \end{aligned}\right\} \tag{3.150}$$

is introduced. In (3.150), $S_{\underline{x}}$, is a positive definite operator which is selfadjoint with respect to the boundary conditions and of the same order as $L_{\underline{x}}$. Under this assumption for $S_{\underline{x}}$, problem (3.150) yields a set of coordinate functions f_i, one corresponding to each Ω_i, which satisfy the same boundary conditions as the solution ϕ of (3.143), and which form a complete system of functions. Moreover, the f_i are orthonormal. Therefore,

$$\left.\begin{aligned} &\int_V f_i(\underline{x}) f_k(\underline{x})\, dV = \delta_{ik}, \\ &\delta_{ik} = \begin{cases} 1 \text{ for } i = k, \\ 0 \text{ for } i \neq k. \end{cases} \end{aligned}\right\} \qquad (3.151)$$

According to [25], solution ϕ of (3.143) can be expanded in terms of the coordinate functions f_i to yield the absolutely and uniformly convergent Fourier series

$$\phi = \sum_{k=1}^{\infty} a_k f_k . \qquad (3.152)$$

Instead of using (3.152), the subsequent operations shall be carried out approximately using the finite series

$$\phi_n = \sum_{k=1}^{n} a_k f_k . \qquad (3.153)$$

Using (3.153) in (3.143) yields

$$\sum_{k=1}^{n} a_k [L_{\underline{x}}(\omega, f_k, \Omega_k) - \omega f_k] = 0. \qquad (3.154)$$

By means of (3.154), the identities

$$\sum_{k=1}^{n} a_k \int_V [L_{\underline{x}}(\omega, f_k, \Omega_k) - \omega f_k] f_i \, dV = 0 \qquad (3.155)$$

can be obtained. Since

$$\int_V [L_{\underline{x}}(\omega, f_k, \Omega_k) - \omega f_k] f_i \, dV = \rho_{ki}(\omega, \Omega_k) \qquad (3.156)$$

is a known quantity, as the f_k are given functions, (3.155) can be rewritten to yield

$$\sum_{k=1}^{n} a_k \rho_{ki}(\omega,\Omega_k) = 0. \tag{3.157}$$

That is a homogeneous system of algebraic equations which serves to determine the coefficients a_k of (3.152). Hence, the stability problem has been algebraicised. Due to (3.157) being homogeneous, nontrivial solutions for the a_k exist only if the condition

$$\det \rho_{ki}(\omega,\Omega_i) = 0, i,k = 1,2,3,\ldots,n, \tag{3.158}$$

is satisfied. As a result of (3.158), the eigenvalue equation

$$\omega_n = \omega_n(\lambda,\Omega_i) \tag{3.159}$$

is obtained where again the structural parameter λ plays the role of a control parameter. Equation (3.159) yields the desired information on stability.

Knowing ω_n allows one to calculate the coefficients $a_k^{(n)}$ from (3.157) and to determine

$$\phi_n = \sum_{k=1}^{n} a_k^{(n)} f_k$$

according to (3.153). All these results are however approximate ones only. For a justification of the preceding calculation scheme, and specifically of Galerkin's method, it has to be shown that ω_n converges towards ω if n in (3.158) is allowed to tend towards infinity. In other words, it has to be shown that the method of reduction applies to the determinant in (3.158). Moreover, it has to be shown that ϕ, to which ϕ_n tends for $n \to \infty$, is a solution of (3.143). This can only be the case if ϕ_n and its derivatives converge absolutely and uniformly. Once that has been established, the proof that (3.143) is being satisfied is possible and will be given later on.

Firstly, let an expansion theorem be derived. Consider a function $g(x)$ which is (2m)-times differentiable and together with all these derivatives continuous. Moreover, $g(x)$ may satisfy the boundary conditions of the auxiliary problem. The operator S_x may be of the order 2m and is supposed to possess all the properties mentioned earlier.

Obviously,

$$S_x(g) \equiv F(x) < K \tag{3.160}$$

is a bounded function in V. Solving (3.160) for g yields

$$g = S_x^{-1}[F(x)]. \tag{3.161}$$

Due to the specific properties of S_x, there exists the bilinearly expandable Green function

$$G(x,\xi) = \sum_{k=1}^{\infty} \frac{f_k(x) f_k(\xi)}{\Omega_k} \tag{3.162}$$

for the auxiliary problem (3.150), and the series on the right hand side of (3.162) converges absolutely and uniformly (see [25] pp. 94-102). Using G yields

$$S_x^{-1} = \int_V G(x,\xi) \ldots d\xi \tag{3.163}$$

so that

$$g(x) = \int_V G(x,\xi) F(\xi) d\xi. \tag{3.164}$$

By virtue of (3.162),

$$g(x) = \int_V \sum_{k=1}^{\infty} \frac{f_k(x) f_k(\xi)}{\Omega_k} F(\xi) d\xi,$$

$$g(x) = \sum_{k=1}^{\infty} \int_V \frac{f_k(\xi) F(\xi)}{\Omega_k} d\xi \, f_k(x), \tag{3.165}$$

respectively, since integration and summation are interchangeable due to the uniform convergence of the series involved. Comparing (3.165) with

$$g(x) = \sum_{k-1}^{\infty} a_k f_k(x) \tag{3.166}$$

proves that

$$a_k = \int_V \frac{f_k(\xi)F(\xi)}{\Omega_k} d\xi. \tag{3.167}$$

Now, it shall be shown, that the series

$$g^{(2m-2-\nu)} = \sum_{k=1}^{\infty} a_k f_k^{(2m-2-\nu)}, \quad \nu = 0,1,2,\dots,2m-2, \tag{3.168}$$

converge uniformly in V. As can be seen easily, it is sufficient to prove the convergence of $g^{(2m-2)}$. If this convergence has been established it implies the convergence of all other series up to the series of g itself as these series can be obtained from $g^{(2m-2)}$ by successive, termwise integration.

Using (3.163) for a transformation of the first line in (3.150) yields

$$f_k(x) = \Omega_k \int_V G(x,\xi) f_k(\xi) d\xi = \int_V G(x,\xi) f_k^{(2m)}(\xi) d\xi, \tag{3.169}$$

which holds true by virtue of $S_x(f_k) \equiv f_k^{(2m)} \equiv \Omega_k f_k$.

Integration by parts and applying the boundary conditions in the second line of (3.150), which are also satisfied by $G(x,\xi)$, transforms (3.169) into

$$f_k(x) = \int_V \frac{\partial^{2m-2} G(x,\xi)}{\partial \xi^{2m-2}} f_k^{(2)}(\xi) d\xi.$$

This relationship together with the first line in (3.150) yields the identity

$$\int_V f_k(x) S_x[f_k(x)] dx = \int_V f_k(x) \Omega_k f_k(x) dx = \int_V \Omega_k f_k(x) \left[\int_V \frac{\partial^{2m-2} G(x,\xi)}{\partial \xi^{2m-2}} f_k^{(2)}(\xi) d\xi \right] dx. \tag{3.170}$$

At the same time,

$$\int_V f_k(x) S_x[f_k(x)]dx = \int_V \left[\sum_{j=1}^{\infty} \delta_{kj} f_j(x) \right] \Omega_k f_k(x)dx. \quad (3.171)$$

By virtue of (3.151), (3.171) changes into

$$\int_V f_k(x) S_x[f_k(x)]dx = \int_V \left[\sum_{j=1}^{\infty} \int_V f_k(\xi) f_j(\xi) d\xi f_j(x) \right] \Omega_k f_k(x)dx. \quad (3.172)$$

By means of (3.150),

$$\int_V f_k(\xi) f_j(\xi) d\xi = \frac{1}{\Omega_j} \int_V f_k(\xi) f_j^{(2m)}(\xi) d\xi$$

is obtained. This relationship can be transformed by integration by parts applied to the right hand side, together with the boundary conditions in (3.150), to yield

$$\int_V f_k(\xi) f_j(\xi) d\xi = \frac{1}{\Omega_j} \int_V f_k^{(2)}(\xi) f_j^{(2m-2)}(\xi) d\xi. \quad (3.173)$$

Using (3.173) in (3.172) leads to

$$\int_V f_k(x) S_x[f_k(x)]dx = \int_V \left[\sum_{j=1}^{\infty} \frac{1}{\Omega_j} \int_V f_k^{(2)}(\xi) f_j^{(2m-2)}(\xi) d\xi f_j(x) \right] \Omega_k f_k(x)dx.$$

This relationship can be changed into

$$\int_V f_k(x) S_x[f_k(x)]dx = \int_V \Omega_k f_k(x) \left[\sum_{j=1}^{\infty} \frac{f_j(x) f_j^{(2m-2)}(\xi)}{\Omega_j} f_k^{(2)}(\xi) d\xi \right] dx. \quad (3.174)$$

Now, compare (3.170) with (3.174). The result is

$$\int_V \Omega_k f_k(x) \left\{ \left[\int_V \frac{\partial^{2m-2} G(x,\xi)}{\partial \xi^{2m-2}} f_k^{(2)}(\xi) d\xi \right] - \left[\sum_{j=1}^{\infty} \int_V \frac{f_j(x) f_j^{(2m-2)}(\xi)}{\Omega_j} f_k^{(2)}(\xi) d\xi \right] \right\} dx = 0. \quad (3.175)$$

This identity can hold true for any x and arbitrary $f_k(x)$ only if

$$\int_V \frac{\partial^{2m-2}G(x,\xi)}{\partial\xi^{2m-2}} f_k^{(2)}(\xi)d\xi = \sum_{j=1}^{\infty} \int_V \frac{f_j(x)f_j^{(2m-2)}(\xi)}{\Omega_j} f_k^{(2)}(\xi)d\xi \qquad (3.176)$$

holds true. Since $G(x,\xi)$ is the Green function of the selfadjoint operator S_x which has the order 2m, $\partial^{2m-2}G/\partial\xi^{2m-2}$ is continuous and therefore bounded. The function $f_k^{(2)}$ is by definition sufficiently smooth and bounded. Hence, the left hand side in (3.176) is bounded. Consequently, the series on the right hand side in (3.176) is uniformly convergent. Therefore, one may interchange summation and integration to obtain

$$\int_V \frac{\partial^{2m-2}G(x,\xi)}{\partial\xi^{2m-2}} f_k^{(2)}(\xi)d\xi = \int_V \sum_{j=1}^{\infty} \frac{f_j(x)f_j^{(2m-2)}(\xi)}{\Omega_j} f_k^{(2)}(\xi)d\xi.$$

From this relationship follows for any x

$$\int_V \left[\frac{\partial^{2m-2}G(x,\xi)}{\partial\xi^{2m-2}} - \sum_{j=1}^{\infty} \frac{f_j(x)f_j^{(2m-2)}(\xi)}{\Omega_j}\right] f_k^{(2)}(\xi)d\xi = 0, \qquad (3.177)$$

which can hold true for any x and ξ and arbitrary $f_k^{(2)}(\xi)$ only if

$$\frac{\partial^{2m-2}G(x,\xi)}{\partial\xi^{2m-2}} = \sum_{j=1}^{\infty} \frac{f_j(x)f_j^{(2m-2)}(\xi)}{\Omega_j}. \qquad (3.178)$$

However, $\partial^{2m-2}G/\partial\xi^{2m-2}$ is bounded for any x and ξ in V, and since (3.178) is supposed to hold true for any x and ξ, the uniform convergence of the series on the right hand side follows from (3.178). Hence,

$$\sum_{j=1}^{\infty} \frac{f_j(x)f_j^{(2m-2)}(\xi)}{\Omega_j} < \tilde{K}. \qquad (3.179)$$

Equation (3.164) serves to derive

$$g^{(2m-2)}(\xi) = \int_V \frac{\partial^{(2m-2)} G(x,\xi)}{\partial \xi^{2m-2}} F(x)dx \tag{3.180}$$

which is a bounded quantity. Moreover, (3.165), (3.166) yield

$$\sum_{k=1}^{n} a_k f_k^{(2m-2)}(\xi) = \sum_{k=1}^{n} \int_V \frac{f_k^{(2m-2)}(\xi) f_k(x)}{\Omega_k} F(x)dx. \tag{3.181}$$

Consequently

$$\left| g^{(2m-2)}(\xi) - \sum_{k=1}^{n} a_k f_k^{(2m-2)}(\xi) \right| = \left| \int_V \left[\frac{\partial^{2m-2} G(x,\xi)}{\partial \xi^{2m-2}} - \sum_{k=1}^{n} \frac{f_k^{(2m-2)}(\xi) f_k(\xi)}{\Omega_k} \right] F(x)dx \right|. \tag{3.182}$$

But because of (3.178),

$$\frac{\partial^{2m-2} G(x,\xi)}{\partial \xi^{2m-2}} - \sum_{k=1}^{n} \frac{f_k^{(2m-2)}(\xi) f_k(x)}{\Omega_k} < \varepsilon.$$

Furthermore, (3.160) holds true. Therefore (3.182) yields

$$\left| g^{(2m-2)}(\xi) - \sum_{k=1}^{n} a_k f_k^{(2m-2)}(\xi) \right| \leqq VK\varepsilon. \tag{3.183}$$

V and K are bounded, ε tends toward zero for increasing n. Hence, (3.183) confirms the uniform convergence of the series

$$g^{(2m-2)}(x) = \sum_{k=1}^{\infty} a_k f_k^{(2m-2)}(x).$$

As already mentioned, this implies also the convergence of the series $g^{(2m-2-\nu)}$, $\nu = 1,2,\dots,2m-2$.

<u>Theorem</u>:

Let g(x) be a (2m)-times continuously differentiable function which satisfies the boundary in (3.150). Then, the series

$$g^{(2m-2-\nu)}(x) = \sum_{k=1}^{\infty} a_k f_k^{(2m-2-\nu)}(x), \quad \nu = 1,2,\ldots,2m-2, \tag{3.184}$$

exist, where the f_k are the eigenfunctions of (2.150), and where the a_k are given by (3.160) and (3.167). These series converge uniformly in V, which is the volume of the body having the boundary B prescribed by the boundary conditions in (3.150). This expansion theorem holds true if the differential operator in (3.150) is of the order 2m and is positive definite and selfadjoint with respect to the boundary conditions.

Now, it shall be assumed that the operator $L_x - \omega E$ (E = unit operator) in (3.143) can be brought into the form

$$L_x - \omega E = S_x - N_x, \tag{3.185}$$

where L_x and S_x are of order 2m, N_x of order 2m-2, and where S_x is positive definite, selfadjoint with respect to the boundary conditions. The operators contained in $\underset{\sim}{U}_x$ (see (3.143)) shall be of an order equal to or smaller than 2m-2.

Because of (3.185), the first line in (3.143) can be rewritten to yield

$$S_x(\phi) = N_x(\omega,\phi). \tag{3.186}$$

For the operator S_x, (3.150) holds true. Also, S_x has the inverse S_x^{-1} which is an integral operator like (3.163). Therefore, (3.186) can be changed into

$$\phi = S_x^{-1}(N_x(\omega,\phi)]. \tag{3.187}$$

Equation (3.187) guarantees the fulfillment of the boundary conditions by ϕ, as the inverse operator S_x^{-1} contains already these boundary conditions.

Now, using the eigenfunctions of (3.150), set up the infinite series (3.152). Substituting this series back in (3.186) yields

$$\sum_{k=1}^{\infty} a_k f_k = S_x^{-1}\left[N_x(\omega, \sum_{k=1}^{\infty} a_k f_k)\right]. \tag{3.188}$$

For ϕ, the expansion theorem holds true: As a solution of (3.143), (3.185), ϕ is a (2m)-times continuously differentiable function if (3.143) possesses continuous coefficients [26]. Since ϕ satisfies the same boundary conditions as they are prescribed in (3.150), this function corresponds exactly to the previously mentioned function g with respect to an expansion in terms of the eigenfunctions f_k of (3.150). Therefore, it can be assumed that the series

$$\phi^{(2m-2-\nu)} = \sum_{k=1}^{\infty} a_k f_k^{(2m-2-\nu)}, \quad \nu = 0,1,2,\ldots,2m-2, \tag{3.189}$$

converge uniformly in V. Since N_x is at most of the order 2m-2, and being a linear operator, this operator can be interchanged with summation by virtue of the uniform convergence of the series involved. The result is

$$\sum_{k=1}^{\infty} a_k f_k = S_x^{-1}\left[\sum_{k=1}^{\infty} a_k N_x(\omega, f_k)\right].$$

But also the operator S_x^{-1} can be interchanged with the summation due to the uniform convergence. Thus, after rearranging,

$$\sum_{k=1}^{\infty} a_k \{S_x^{-1}[N_x(\omega,f_k)] - f_k\} = 0 \tag{3.190}$$

is obtained. This relationship may be used to derive the identity

$$\sum_{k=1}^{\infty} a_k \int_V \{S_x^{-1}[N_x(\omega,f_k)] - f_k\} f_i dV = 0 \tag{3.191}$$

which by means of

$$\int_V \{S_x^{-1}[N_x(\omega,f_k)] - f_k\}f_i dV = \tilde{\rho}_{ki} \tag{3.192}$$

yields

$$\sum_{k=1}^{\infty} a_k \tilde{\rho}_{ki} = 0. \tag{3.193}$$

This infinite system of equations is assumed to be solvable by means of the <u>method of reduction</u>. Hence, the a_k can be evaluated, and the series (3.189) are known.

It shall be shown that (3.189) yields the solution of (3.143) under the assumption that (3.185) is valid. By virtue of the uniform convergence, (3.191) can be changed into

$$\int_V \left[\sum_{k=1}^{\infty} a_k\{S_x^{-1}[N_x(\omega,f_k)] - f_k\}f_i dV\right] = 0.$$

Since f_k, k = 1,2,3,..., is a complete system of functions, the preceding relationship implies

$$\sum_{k=1}^{\infty} a_k\{S_x^{-1}[N_x(\omega,f_k)] - f_k\} = 0$$

or

$$\sum_{k=1}^{\infty} a_k f_k = \sum_{k+1}^{\infty} a_k S_x^{-1}[N_x(\omega,f_k)].$$

But operators S_x^{-1} and N_x are interchangeable with the summation by virtue of uniform convergence. Therefore,

$$\sum_{k=1}^{\infty} a_k f_k = S_x^{-1} N_x(\omega, \sum_{k=1}^{\infty} a_k f_k)$$

results. Applying to both sides the operator S_x yields

$$S_x\left(\sum_{k=1}^{\infty} a_k f_k\right) = N_x\left(\omega, \sum_{k=1}^{\infty} a_k f_k\right),$$

$$S_x\left(\sum_{k=1}^{\infty} a_k f_k\right) - N_x\left(\omega, \sum_{k=1}^{\infty} a_k f_k\right) = 0,$$

respectively. Using (3.185) yields finally

$$L_x\left(\sum_{k=1}^{\infty} a_k f_k\right) - \omega\left(\sum_{k=1}^{\infty} a_k f_k\right) = 0,$$

which proves that (3.189) is indeed the solution of (3.143). The boundary conditions in (3.143) are satisfied by (3.189) without saying as the coordinate functions f_k satisfy these boundary conditions by definition.

The previous considerations are based on the assumption that the method of reduction is applicable to the system of equations (3.193). Let it be shown under which circumstances that is in fact the case. According to H. V. Koch [27] and F. Riesz [28], the following conditions must be satisfied:

$$\sum_{i=1}^{\infty} |\tilde{\rho}_{ii} + 1| \text{ convergent,} \tag{3.194}$$

$$\sum_{i=1}^{\infty} \sum_{k=1}^{\infty} |\tilde{\rho}_{ki} + \delta_{ki}|^2 \text{ convergent,} \tag{3.195}$$

$$\sum_{i=1}^{\infty} a_i^2 \text{ convergent.} \tag{3.196}$$

According to (3.192) and (3.151),

$$\tilde{\rho}_{ki} = \int_V S_x^{-1}[N_x(\omega, f_k)] f_i dV - \delta_{ki}.$$

By means of (3.163), this expression is changed into

$$\tilde{\rho}_{ki} = \int_V \int_V G(x,\xi)\,[N_x(\omega, f_k(\xi))]\, f_k(x)\,dx\,d\xi - \delta_{ki}. \tag{3.197}$$

But $G(x,\xi)$ is expandable according to (3.162), and that expansion is absolutely and uniformly convergent. Therefore, integration and summation can be interchanged, and by virtue of (3.151),

$$\int_V G(x,\xi) f_k(x)\,dx = \sum_j \int_V \frac{f_j(x) f_j(\xi)}{\Omega_j} f_k(x)\,dx = \frac{f_k(\xi)}{\Omega_k}. \tag{3.198}$$

Using (3.198) in (3.197) yields

$$\tilde{\rho}_{ki} = \int_V \frac{N_x(\omega, f_k(\xi))}{\Omega_i} f_i(\xi)\,d\xi - \delta_{ki}. \tag{3.199}$$

Since N_x is supposed to be a linear operator of the order 2m-2,

$$N_x[\omega, f_k(\xi)] = \sum_{i=0}^{2m-2} g_i(\xi) f_k^{(i)}(\xi), \tag{3.200}$$

where the $g_i(\xi)$ are sufficiently smooth functions. By means of (3.200), (3.199) can be rewritten to yield

$$\tilde{\rho}_{ki} = \frac{1}{\Omega_i} \sum_{j=0}^{2m-2} \left[\int_V g_j(\xi) f_k^{(j)}(\xi) f_i(\xi)\,d\xi\right] - \delta_{ki}. \tag{3.201}$$

Introducing the notation

$$\beta_{ki} = \sum_{j=0}^{2m-2} \left[\int_V g_j(\xi) f_k^{(j)}(\xi) f_i(\xi)\,d\xi\right], \tag{3.202}$$

(3.201) can be transformed into

$$\tilde{\rho}_{ki} = \frac{\beta_{ki}}{\Omega_i} - \delta_{ki}. \tag{3.203}$$

Using (3.203) in (3.194) yields the condition that

$$\sum_{i=1}^{\infty}\left|\frac{\beta_{ii}}{\Omega_i}\right| = \sum_{i=1}^{\infty}\left|\frac{1}{\Omega_i}\sum_{j=0}^{2m-2}\int_V g_j(\xi)f_i^{(j)}(\xi)f_i(\xi)d\xi\right| \text{ be convergent.} \tag{3.204}$$

But

$$\sum_{i=1}^{\infty}\left|\frac{1}{\Omega_i}\sum_{j=0}^{2m-2}\int_V g_j(\xi)f_i^{(j)}(\xi)f_i(\xi)d\xi\right| \leqq \sum_{j=0}^{2m-2}\sum_{i=1}^{\infty}\left|\frac{\int_V g_j(\xi)f_i^{(j)}(\xi)f_i(\xi)d\xi}{\Omega_i}\right| . \tag{3.205}$$

Therefore,

$$\sum_{i=1}^{\infty}\left|\frac{\beta_{ii}}{\Omega_i}\right| \leqq \sum_{j=0}^{2m-2}\sum_{i=1}^{\infty}\left|\frac{\int_V g_j(\xi)f_i^{(j)}(\xi)f_i(\xi)d\xi}{\Omega_i}\right| . \tag{3.206}$$

Assume that $g_j(\xi)f_i^{(j)}(\xi)$ is a semi-permissible function with respect to (3.150) in the sense of L. Collatz [25, pp. 141, 142]. That is the case if $g_j(\xi)f_i^{(j)}(\xi)$ is (2m)-times differentiable, i.e. $g_j(\xi)$ must be in the class C_{2m} and f_i must be (2m-2+2m) = (4m-2)-times differentiable. Then, the equivalence

$$g_j(\xi)f_i^{(j)}(\xi) \sim \sum_{s=1}^{\infty} \alpha_{si}^{j} f_s(\xi) \tag{3.207}$$

holds true. In (3.207), the

$$\alpha_{si}^{j} = \int_V g_j(\xi)f_i^{(j)}(\xi)f_s(\xi)d\xi \tag{3.208}$$

are the Fourier coefficients of the expansion (3.207), for which the inequality of Parseval,

$$\sum_{s=1}^{\infty} (\alpha_{si}^{j})^2 \leqq \int_V g_j^2(\xi)[f_i^{(j)}(\xi)]^2 d\xi, \tag{3.209}$$

holds true. Since the term on the right hand side of (3.209) is bounded, the quantities $|\alpha_{si}^{j}|$ are the elements in the table

$$\begin{array}{ccccc}
|\alpha_{11}{}^{j}| & |\alpha_{12}{}^{j}| & |\alpha_{13}{}^{j}| & |\alpha_{14}{}^{j}| & \ldots \\
|\alpha_{21}{}^{j}| & |\alpha_{22}{}^{j}| & |\alpha_{23}{}^{j}| & |\alpha_{24}{}^{j}| & \ldots \\
|\alpha_{31}{}^{j}| & |\alpha_{32}{}^{j}| & |\alpha_{33}{}^{j}| & |\alpha_{34}{}^{j}| & \ldots \\
\vdots & & & \ddots &
\end{array}$$

for which any line represents necessarily a null sequence. The quantities $|\alpha_{ii}{}^{j}|$, i = 1,2,3,... are the diagonal elements in this table and therefore bounded in their magnitude. For example, $|\alpha_{ii}{}^{j}| < M$ for any i and j. Using this inequality and in addition (3.208) in (3.206) yields

$$\sum_{i=1}^{\infty}\left|\frac{\beta_{ii}}{\Omega_i}\right| \leqq \sum_{j=0}^{2m-2}\sum_{i=1}^{\infty}\left|\frac{\alpha_{ii}{}^{j}}{\Omega_i}\right| \leqq \sum_{i=1}^{\infty}\frac{2m-1}{\Omega_i}M = (2m-1)M\sum_{i=1}^{\infty}\frac{1}{\Omega_i}. \tag{3.210}$$

However, the eigenvalues Ω_i of the auxiliary problem (3.150) satisfy the relationship

$$\sum_{i=1}^{\infty}\frac{1}{\Omega_i} = \int_V G(x,x)\,dx = D, \tag{3.211}$$

where D is a constant. That is true because of (3.163) and the specific properties of S_x. Using (3.211) in (3.210) and observing (3.194), (3.203),

$$\sum_{i=1}^{\infty}\left|\tilde{\rho}_{ii} + 1\right| = \sum_{i=1}^{\infty}\left|\frac{\beta_{ii}}{\Omega_i}\right| \leqq (2m-1)MD \tag{3.212}$$

results which means convergence. Hence, condition (3.194) is satisfied.

Now, let condition (3.195) be considered. First,

$$\sum_{k=1}^{\infty}\left|\tilde{\rho}_{ki} + \delta_{ki}\right|^2 = \sum_{k=1}^{\infty}\left|\frac{\beta_{ki}}{\Omega_i}\right|^2 \tag{3.213}$$

$$\sum_{k=1}^{\infty}\left(\frac{\beta_{ki}}{\Omega_i}\right)^2 = \left[\sum_{k=1}^{\infty}\sum_{j=0}^{2m-2}\int_V g_j(\xi) f_i^{(j)}(\xi) f_k(\xi) d\xi\right]^2 \cdot \frac{1}{\Omega_1^{\,2}} = \sum_{k=1}^{\infty}\left[\sum_{j=0}^{2m-2} \alpha_{ki}^{j}\right]^2 \frac{1}{\Omega_1^{\,2}} .$$

By means of Canchy's inequality, that is to say

$$(a_1 + a_2 + a_3 + \ldots + a_n)^2 \leqq n(a_2^{\,2} + a_2^{\,2} + a_2^{\,2} + \ldots + a_n^{\,2}),$$

the previous relationship can be changed into the inequality

$$\sum_{k=1}^{\infty}\left(\frac{\beta_{ki}}{\Omega_i}\right)^2 \leqq \frac{1}{\Omega_i^{\,2}} (2m-1) \sum_{i=1}^{\infty}\sum_{j-0}^{2m-2} (\alpha_{ki}^{j})^2 .$$

The series on the right hand side is unconditionally convergent and can therefore be rearranged. Therefore, using also (3.209),

$$\sum_{k=1}^{\infty}\left(\frac{\beta_{ki}}{\Omega_i}\right)^2 \leqq \frac{2m-1}{\Omega_i^{\,2}} \sum_{j=0}^{2m-2}\sum_{i=1}^{\infty} (\alpha_{ki}^{j})^2 \leqq \frac{2m-1}{\Omega_i^{\,2}} \sum_{j=0}^{2m-2}\int_V g_j^{\,2}(\xi)\,[f_i^{(j)}(\xi)]^2 d\xi \tag{3.214}$$

is obtained. Since the functions g_j are by definition in class C_{2m}, the mean value theorem of integration can be applied.

$$\sum_{k=1}^{\infty}\left(\frac{\beta_{ki}}{\Omega_i}\right)^2 \geqq \frac{(2m-1)\widetilde{M}}{\Omega_i^{\,2}} \sum_{j=0}^{2m-2}\int [f_i^{(j)}(\xi)]^2 d\xi \tag{3.215}$$

results, where $\widetilde{M}$ is an appropriate constant. Therefore, (3.213) and (3.215) yield

$$\sum_{i=1}^{\infty}\sum_{k=1}^{\infty}\left|\tilde{\rho}_{ki} + \delta_{ki}\right|^2 \leqq (2m-1)\widetilde{M} \sum_{i=1}^{\infty}\sum_{j=0}^{2m-2} \frac{\int_V [f_i^{(j)}(\xi)]^2 d\xi}{\Omega_i^{\,2}} .$$

Since a rearrangement of the series is again permitted,

$$\sum_{i=1}^{\infty}\sum_{k=1}^{\infty}\left|\tilde{\rho}_{ki} + \delta_{ki}\right|^2 \leqq (2m-1)\widetilde{M} \sum_{j=0}^{2m-2}\sum_{i=1}^{\infty} \frac{\int_V [f_i^{(j)}(\xi)]^2 d\xi}{\Omega_i^{\,2}} \tag{3.216}$$

is obtained.

From (3.150) and (3.163) follows

$$f_k(x) = \Omega_k \int_V G(x,\xi) f_k(\xi) d\xi.$$

Therefore,

$$f_k^{(j)}(x) = \Omega_k \int_V \frac{\partial^{(j)} G(x,\xi)}{\partial x^j} f_k(\xi) d\xi.$$

The equivalence

$$\frac{\partial^{(j)} G(x,\xi)}{\partial x^j} \quad \sum_k \frac{f_k^{(j)}(x) f_k(\xi)}{\Omega_k}$$

yields

$$\frac{f_k^{(j)}(x)}{\Omega_k} = \int_V \frac{\partial^{(j)} G(x,\xi)}{\partial x^j} f_k(\xi) d\xi$$

as the Fourier coefficient of this equivalence. Therefore, the inequality of Parseval

$$\sum_{k=1}^{\infty} \frac{[f_k^{(j)}(x)]^2}{\Omega_k^2} \leqq \int_V \left[\frac{\partial^{(j)} G(x,\xi)}{\partial x^j}\right]^2 d\xi = F_j(x) \tag{3.217}$$

holds true. Since $\partial^j G/\partial x^j$ is continuous for $j \leqq 2m-2$, also the functions $F_j(x)$ are continuous. Therefore, the series

$$\sum_{j=1}^{\infty} \frac{[f_k^{(j)}(x)]^2}{\Omega_k^2} \tag{3.218}$$

converge uniformly in V for $j \leqq 2m-2$.

Because of (3.217) and the uniform convergence of the series (3.218)

$$\sum_{i=1}^{\infty} \int_V \frac{[f_i^{(j)}(\xi)]^2}{\Omega_i^2} d\xi = \int_V \sum_{i=1}^{\infty} \frac{[f_i^{(j)}(\xi)]^2}{\Omega_i^2} d\xi \leqq \int_V F_j(\xi) d\xi = \approx{M}_j. \tag{3.219}$$

This inequality is used in (3.216). The result is

$$\sum_{i=1}^{\infty}\sum_{k=1}^{\infty}\left|\tilde{\rho}_{ki}+\delta_{ki}\right|^2 \leqq (2m-1)\,\tilde{M}\sum_{j=0}^{2m-2}\approx{M}_j .$$

On the right hand side there is a constant. Hence, the series on the left hand side is convergent. That proves that also condition (3.195) is satisfied.

Condition (3.196) reads

$$\sum_{i=1}^{\infty} a_i^2 = \sum_{i=1}^{\infty}\left[\int_V \frac{f_i(\xi)F(\xi)}{\Omega_i}\,d\xi\right]^2, \text{ bounded,}$$

when using (3.167). Applying Schwarz' inequality yields first

$$\sum_{i=1}^{\infty} a_i^2 \leqq \sum_{i=1}^{\infty}\left[\int_V \frac{[f_i(\xi)]^2}{\Omega_i^2}\,d\xi \,.\, \int_V F^2(\xi)d\xi\right]$$

and then, by virtue of the boundedness of F (see (3.160)),

$$\sum_{i=1}^{\infty} a_i^2 \leqq M^* \sum_{i=1}^{\infty}\int_V \frac{[f_i(\xi)]^2}{\Omega_i^2}\,d\xi, \qquad (3.220)$$

where

$$M^* = \int_V F^2(\xi)d\xi.$$

However,

$$\sum_{i=1}^{\infty}\frac{[f_i(\xi)]^2}{\Omega_i^2} \leqq \int_V [G(x,\xi)]^2 dx, \qquad (3.221)$$

which follows from (3.217) for j = 0. Since the series on the left hand side converges uniformly, (3.221) can be used to yield

$$\sum_{i=1}^{\infty}\int_V \frac{[f_i(\xi)]^2}{\Omega_i^2}\,d\xi = \int_V \sum_{i=1}^{\infty}\frac{[f_i(\xi)]^2}{\Omega_i^2}\,d\xi \leqq \int_V\int_V [G(x,\xi)]^2 dx d\xi. \qquad (3.222)$$

Using (3.222) in (3.220) yields

$$\sum_{i=1}^{\infty} a_i^2 \leq M^* \int_V \int_V [G(x,\xi)]^2 dx d\xi$$

which means convergence. Hence, condition (3.196) is satisfied.

The conclusion is that the method of reduction can indeed be applied to (3.193). Therefore, the coefficients a_k of the solution (3.189) to (3.143) can be determined. Consequently, this solution can definitely be constructed.

The preceding results can be summarized in the following theorem: Let the method of Galerkin be applied to (3.143). If for the differential operators in (3.143), condition (3.185) holds true, i.e.

$$L_x - \omega E = S_x - N_x,$$

where S_x is of order 2m, linear, positive definite, selfadjoint with respect to the boundary conditions in (3.143),

N_x is of order 2m-2 and has the structure

$$N_x = \sum_{j=0}^{2m-2} g_j f_i^{(j)},$$

g_i is in the Class C_{2m},

and if the coordinate function f_k used in Galerkin's method are the eigenfunctions of the auxiliary problem (3.150) and are (4m-2)-times differentiable, then the method of Galerkin yields by means of (3.193) the coefficients a_k which are needed to set up the uniformly convergent series (3.189). The series

$$\phi = \sum_k a_k f_k,$$

which is contained in (3.189) for ν = 2m-2, is the exact solution to (3.186) which is a problem equivalent to the original problem (3.143) from which it can be derived by means of a simple transformation.

4. Conclusion

It has been shown that the investigation of the stability of an elastic system can be reduced under fairly general conditions mathematically to the treatment of a so called fundamental problem. This is a homogeneous variational equation and a set of boundary conditions.

Next, attention has been paid how to determine eigenvalues and eigenfunctions of the fundamental problem as these quantities govern stability according to their behaviour. Three approaches to the solution have been presented: Firstly, Liapunov's method in an extended version due to Movchan and others. This is a direct method applicable to the differential equation and yielding stability criteria and conditions. Secondly, the energy method. If the work done by non-conservative forces is included, this method is generally applicable. In the classical conservative case, it is reduced to a discussion of the potential energy surface, and ample use of it has been made in Part II of this monograph in that reduced from. Thirdly, the modal approach, together with Galerkin's method, has been presented. This approach is very common and has widely been applied in Part II and III of the monograph. It leads to a discretization and algebraization of the fundamental problem. In Part II, frequently, the starting point of the discussions was immediately the algebraic version of the fundamental problem. Therefore, specific emphasis was put in Part I on the justification of that approach. It depends on the admissibility of series expansions whose convergence has been proved and then, on the reducibility of an infinite matrix. Also, the question of reducibility has been takled thoroughly.

It is hoped that the investigation contained in Part I rapresents a solid mathematical foundation for the work presented in the subsequent Parts II and III.

Figures, Part I

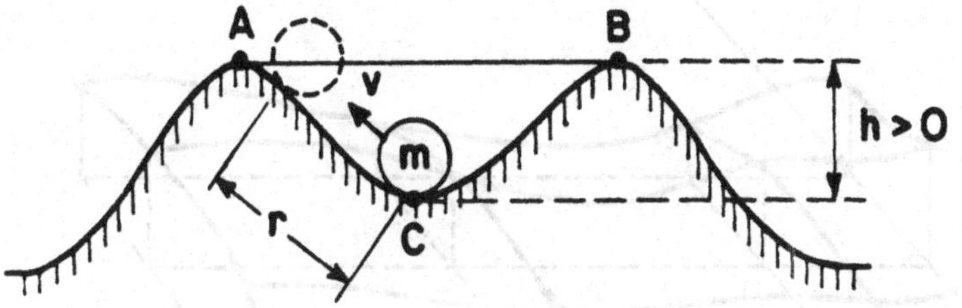

(a) Position C stable, degree of stability non-zero, positive

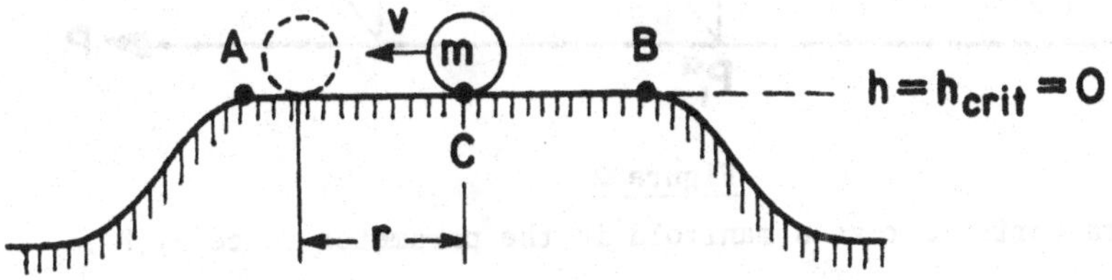

(b) Stability Limit, degree of stability vanished

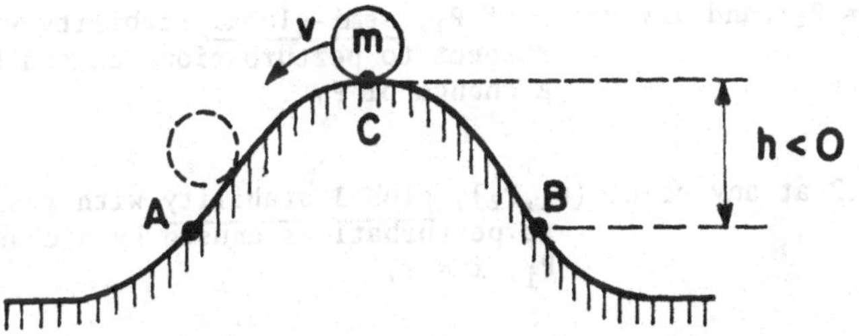

(c) Instability, degree of stability negative

Figure 1

Example of a Stability Problem

h = control parameter, h_{crit} = critical value of control parameter,
r = characteristic,
I = mv = perturbation,
m = mass of ball, v = velocity of ball.

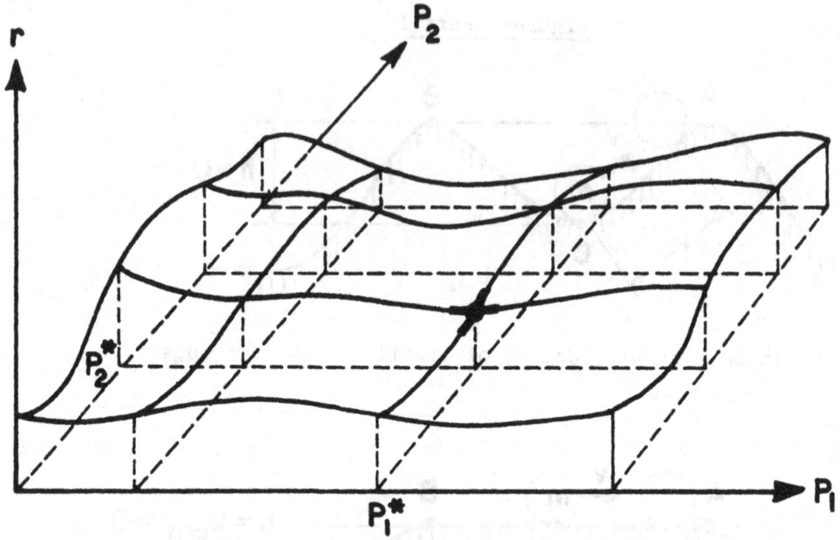

Figure 2

The characteristic r as a manifold in the parameter space P_1,P_2.

$\left|\frac{\partial r}{\partial P_i}\right| = c < \infty$, $i = 1,2$, at (P_1^*,P_2^*), local stability with respect to perturbations caused by a change of P_i, $i = 1,2$

$\left|\frac{\partial r}{\partial P_1}\right| = c < \infty$ at $P_2 = P_2^*$ and any value of P_1, semi-global stability with respect to perturbations caused by a change of P_1

$\left|\frac{\partial r}{\partial P_i}\right| = c < \infty$, $i = 1,2$ at any point (P_1,P_2), global stability with respect to perturbations caused by a change of P_i, $i = 1,2$

$\left|\frac{\partial r}{\partial P_2}\right| = c < \infty$ at $P_2 = P_2^*$ and any value of P_1, increased Liapunov stability with respect to perturbations caused by a change of P_2

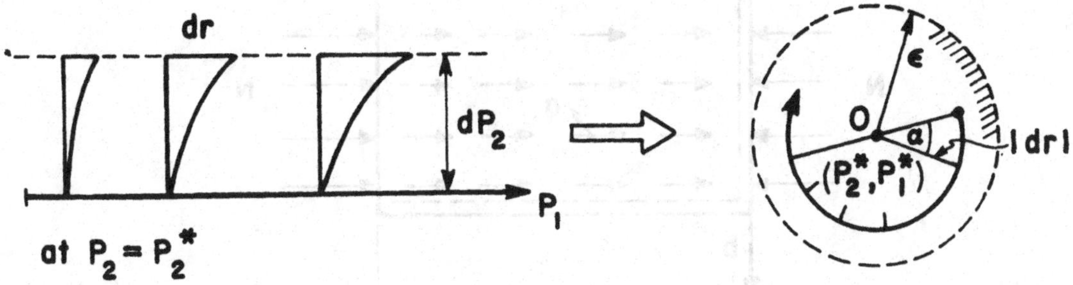

α = arc radian $|P_1|$

ε = $|dP_2|$.c, c = const.

Figure 3

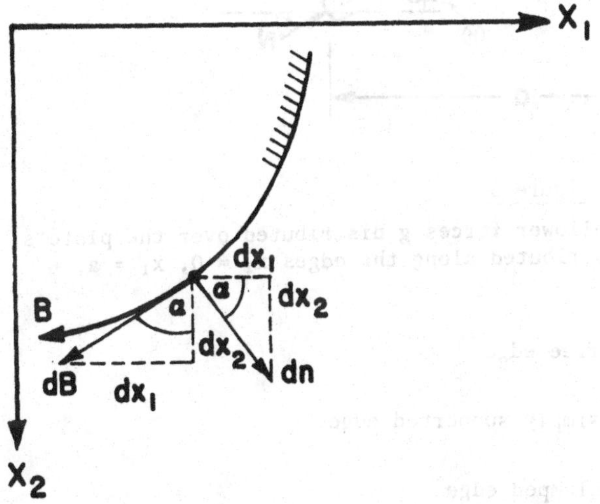

Figure 4

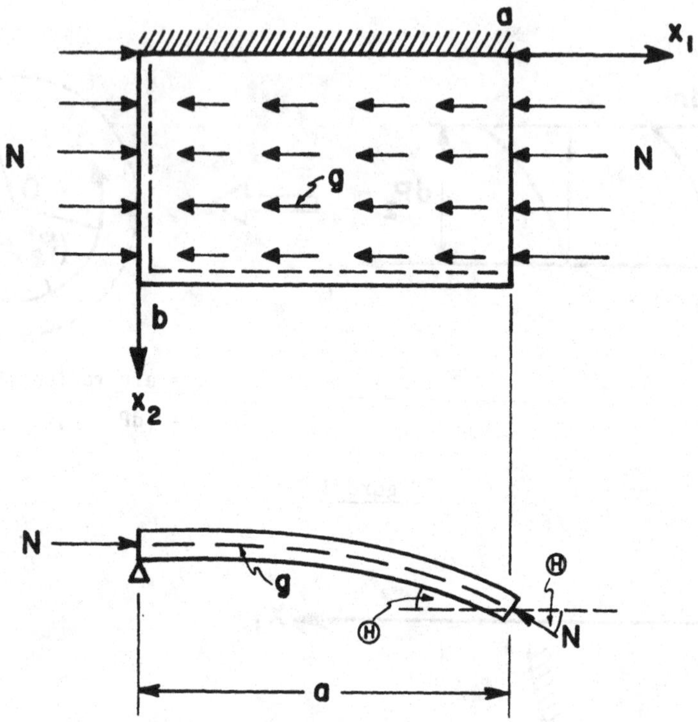

Figure 5

Rectangular plate subjected to follower forces g distributed over the plate's surface and follower forces N distributed along the edges $x_1 = 0$, $x_1 = a$.

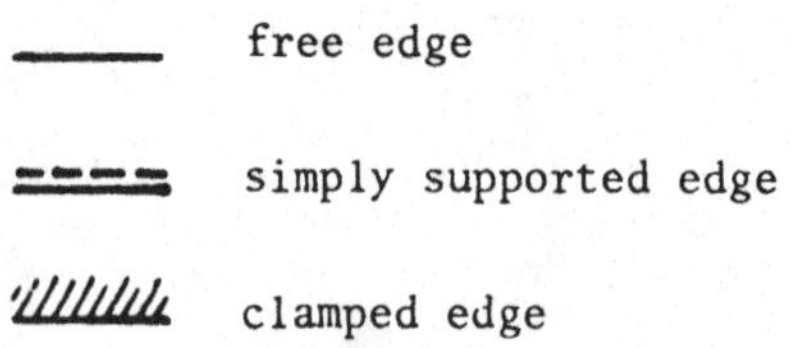

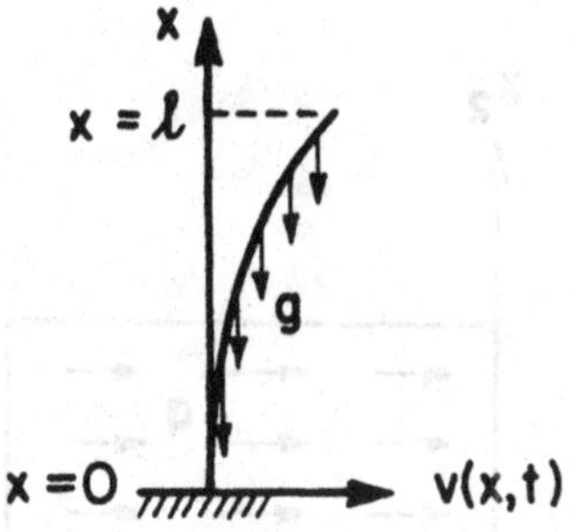

Figure 6
Elastic Rod Subjected to Uniformly Distributed Compression

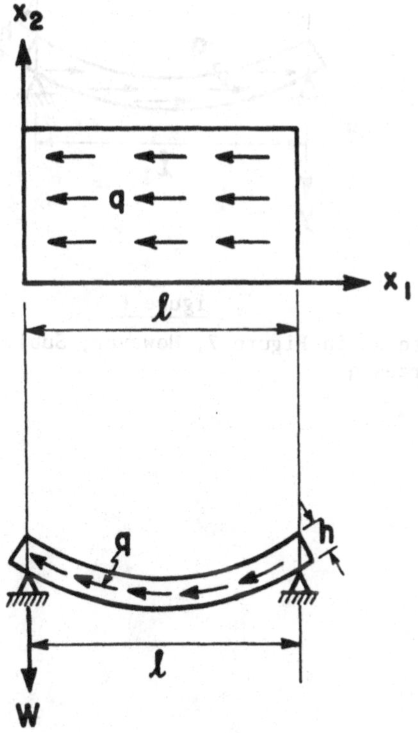

Figure 7
Plate with Constant Thickness h Subjected to Nonconservative Follower Forces q

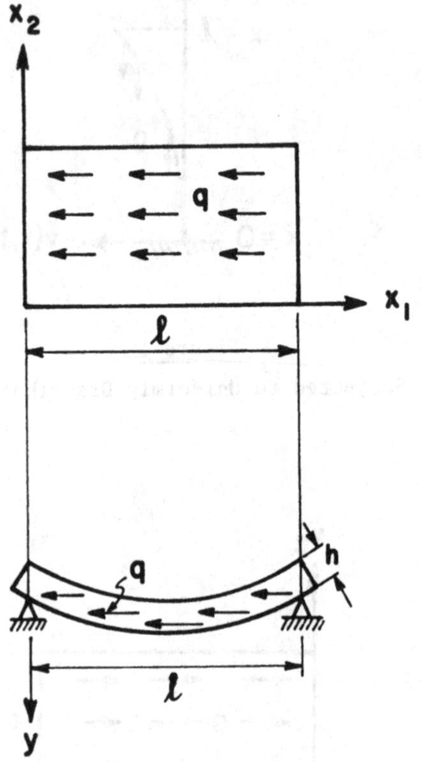

Figure 8

Corresponding Plate as in Figure 7, However, Subjected to Conservative Unidirectional Forces q

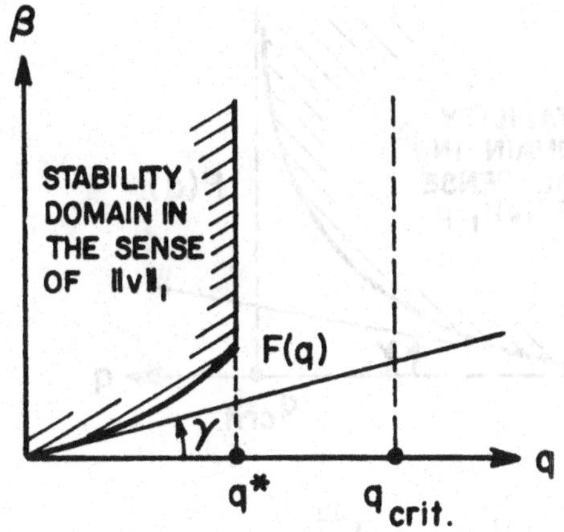

$$\tan\gamma = \left(\frac{\mu}{mq_{crit.}}\right)^{1/2}$$

$$F(q) = \frac{q\sqrt{\mu}}{[m(q_{crit.}-q]^{1/2}}$$

$$q^* = \frac{D(1-\sigma)}{mA}$$

$$||v||_1 = \sup|v|$$

Figure 9

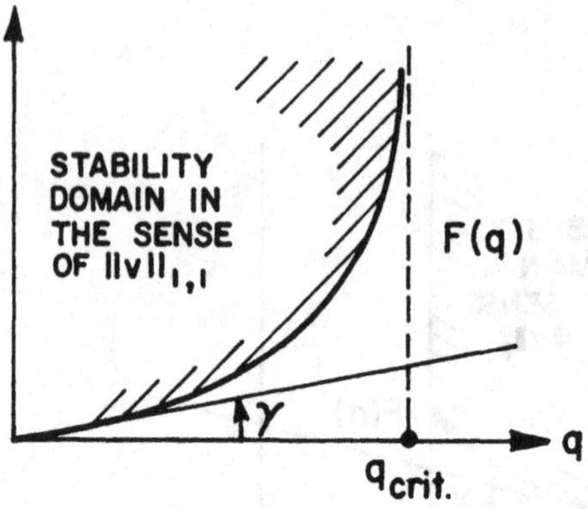

$$\tan \gamma = \left(\frac{\mu}{m q_{crit.}}\right)^{1/2}$$

$$F(q) = \frac{q\sqrt{\mu}}{[m(q_{crit.} - q]^{1/2}}$$

$$||v||_{1,1} = [\int_s v^2 dS]^{1/2}$$

Figure 10

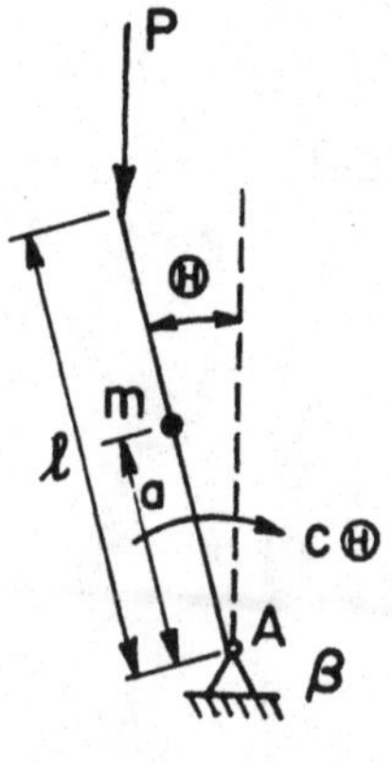

Figure 11

References, Part I

[1] LEIPHOLZ, H. H. E., "On Destabilizing Parameters, the Degree of Stability of Mechanical Systems and General Stability Criteria", Transactions of CSME 3, 1975, pp. 103-110.

[2] LEIPHOLZ, H. H. E., "Stabilitat Elastischer Systeme", Braun-Verlag, Karlsruhe, 1978.

[3] LEIPHOLZ, H., "Stability Theory", Academic Press, New York and London, 1970, pp. 134-139.

[4] SCHRÄPEL, H. D., "Ein Beitrag zur ersten Liapunowschen Methode", ZAMM53,T243, 244, 1973.

[5] LEIPHOLZ, H, "Direct Variational Methods and Eigenvalue Problems in Engineering", Noordhoff International Publ., Leyden, 1977, pp. 178-182.

[6] LIAPUNOV, A. M., "Problème Générale de la Stabilite du Mouvement", Am. Fac. Sci., Toulouse, 9, 1907, pp. 203-474.

[7] MALKIN, J. G., "Theory of Stability of Motion", AEC-tr-3352, Office of Technical Services, Department of Commerce, Washington, D.C., U.S.A.

[8] CHETAYEV, N. G., "The Stability of Motion", Pergamon Press, Oxford, London, New York, Paris, 1961.

[9] HAHN, W., "Theory and Application of Liapunov's Direct Method", Prentice-Hall, Inc., Englewood Cliffs, N.J., 1963.

[10] ZUBOV, V. I., "The Methods of Liapunov and Their Applications", Leningrad, 1957.

[11] MOVCHAN, A. A., "The Direct Method of Liapunov in Stability Problems of Elastic Systems", J. Appl. Math. Mech., 23, pp. 670-686, "Stability of Processes with Respect to Two Metrics", J. Appl. Math. Mech., 24, pp. 1506-1524.

[12] KNOPS, R. J. and WILKES, E. W., "On Movchan's Theorems for Stability of Continuous Systems', Int. J. Eng. Sci., 4, pp. 303-329.

[13] KNOPS, R. J. and WILKES, E. W., "Theory of Elastic Stability" Encyclopedia of Physics, Vol. VI a/3, Mechanics of Solids III, pp. 125-302, S. Pflüger, Chief Editor, C. Truesdell, Editor, Springer-Verlag, Berlin, Heidelberg, New York, 1973.

[14] DYM, C. L., "Stability Theory and Its Applications to Structural Mechanics", Noordhoff International Publ., Leyden, 1974.

[15] PARKS, P. C., "A Stability Criterion for Panel Flutter Problem via the Second Method of Liapunov", AIAA Journal, 3, No. 9, 1965, pp. 1764-1766.

[16] LEIPHOLZ, H. H. E., "Application of Liapunov's Direct Method to the Stability Problem of Rods Subjected to Follower Forces", "Instability of Continuous Systems", Springer-Verlag, Berlin, Heidelberg, New York, 1971, pp. 1-10.

Über die Anwendung von Liapunov's direkter Methode auf Stabilitätsprobleme kontinuierlicher, nichtkonservativer Systems, Ing. Arch., 39, 1970, pp. 257-268.

On Conservative Elastic Systems of the First and Second Kind, Ing. Arch., 43, 1974, pp. 255-271.

Some Remarks on Liapunov Stability of Elastic Dynamical Systems, Buckling of Structures, IUTAM Symposium, Cambridge, Ma., U.S.A., 1974, B. Budiansky, Editor, Springer-Verlag, Berlin, Heidelberg, New York, 1976.

Stability of Elastic Rods via Liapunov's Second Method, Ing. Arch., 44, 1975, pp. 21-26.

Stability of Elastic Plates via Liapunov's Second Method, Ing. Arch., 45, 1976, pp. 337-345.

[17] BROWN, A. L. and PAGE, A., "Elements of Functional Analysis", Van Nostrand Reinhold Co., London, New York, Cincinnati, Toronto, Melbourne, 1970, pp. 85-90.

[18] GOLDSTEIN, H., "Classical Mechanics", Addison-Wesley Publ. Co., Reading, Massachusetts, Palo Alto, London, Dallas, Atlanta, 7th Printing, 1965, pp. 359-364.

[19] FRANK, Ph. and MISES, R. V., "Die Differential -und Integralgleichungen der Mechanik und Physik", Band II, pp. 129-133, Fr. Vieweg and Son, Braunschweig, 1934.

[20] LEIPHOLZ, H., "Stability Theory", Academic Press, New York, London, 1970, p. 188.

[21] ROTHE, R. and SZABO, I., "Hohere Mathematik", Teil VI, pp. 172, 173, B. G. Teubner, Stuttgart, 1953.

[22] LEIPHOLZ, H. H. E., "Liapunov's Second Method and Its Application to Continuous Systems", SM Arch. 1, 1976, pp. 367-444.

[23] ZIEGLER, H., Ing. Arch., 20, 1952, p. 49.

[24] LEIPHOLZ, H. H. E., "On the Application of the Energy Method to the Stability Problem of Nonconservative Autonomous and Nonautonomous Systems", Acta Mechanica, to appear.

[25] COLLATZ, L., "Eigenwertaufgaben mit technischen Anwendungen", Akad. Verlagsges, Geest u. Portig K.-G., Leipzig, 1963, 2. Aufl., pp. 137-149.

[26] HORN, I., "Gewöhnliche Differentialgleichungen", Walter de Gruyter et Co., Berlin W35, 1948, 5-te Aufl. pp. 56, 57.

[27] KOCH, H. V., "Sur la convergence des déterminants infinis", Rendiconti del Circolo mat. di Palermo, t. XXVIII, 1909, p. 255.

[28] RIESZ, F., "Les systèmes d'équations linéaires à une infinite d'inconnues", Collection Borel, Paris, 1913.

PART II

by K. Huseyin
University of Waterloo),
Ontario, Canada

Introduction

This part is concerned with the stability of discrete systems whose behaviour is influenced by several independent parameters, and it will be presented in two subsequent chapters.

If a system possesses a potential (gradient system), such as the potential energy associated with conservative problems, the stability behaviour can be studied conveniently via statical methods, and the first chapter of this article will be devoted to multiple parameter gradient systems. In the case of non-gradient systems, however, stability investigations often require dynamical criteria. In fact, Liapunov's definition of stability is a dynamical one, and it is only natural to employ dynamical methods in most general cases. The second chapter will, therefore, be concerned with dynamical behaviour and stability of multiple-parameter systems. In chapter 1, a general nonlinear theory will be developed while in chapter 2, the treatment will be confined to general linear developments.

1. Stability of Gradient Systems

1.1 Introductory remarks

It is well-known that the loss of stability of a gradient system, whose behaviour is influenced by only one independent parameter, can be associated either with a bifurcation or a limit point. It may be argued that the former can occur in our idealized mathematical models while the nature with all its imperfections normally dictates the latter. Nevertheless, the practical as well as theoretical value of the concept of bifurcation in stability theory is unquestionable. In fact, the studies concerning the effect of imperfections on the stability behaviour of structural systems are based on the bifurcation points associated with a perfect model of the system. Thus, in the development of the nonlinear theory of elastic stability in Mechanics, bifurcation and limit points played a central role along the analyses concerning imperfections essentially involve at least two parameters (a load and an imperfection parameter [1, 2, 3]). As a matter of fact, systems are often under the influence of several parameters (this is a rule rather than an exception in our environment), and it was first pointed out in reference [4] that bifurcation and limit points are not quite adequate to describe the behaviour of multiple-parameter systems fully, and a re-classiciation of critical conditions based on equilibrium surfaces were introduced. In this classification, proper surfaces are distinguished from improper (degenerate) surfaces and mainly two types of critical points, namely <u>general</u> and <u>special</u> points, describing the former and latter situation respectively, were introduced. A general nonlinear multiple-parameter theory of stability was then developed on the basis of these new concepts [5]. Catastrophe theory [6] essentially parallels this stability theory, and provides a full qualitative classification of instability phenomena (catastrophes) which can arise when the system has four parameters. All the catastrophes, however, are associated with smooth surfaces and hence with <u>general</u> critical points. A discussion concerning the connection between the two theories is presented in reference [7]. While it is true that the nature, with all its imperfections, will normally produce <u>general</u> points, the analytical value of special points, nevertheless, is enormous as pointed out earlier. An example may be cited to illustrate this point. Consider a flat thin plate subjected to two independent uniformly distributed axial compression forces along its perpendicular edges. It is expected, of course, that this system will have small imperfections. However, an idealized model which initially ignores such imperfections is invaluable in a quantitative analysis. In fact, the most convenient method of analysis concerning the effect of small imperfections is based on the perfect model whose loss of stability is associated with special critical points. A thorough knowledge pertaining to this phenomenon is, therefore, almost a prerequisite for exploring the behaviour of the imperfect system, and not the other way round.

On the other hand, the multiplicity of eigenvalues associated with critical points play an important role on the behaviour of the system and may present some complications. It is the main objective of this article to present a general discussion without restriction in this regard.

1.2 Classification of critical conditions

Consider a system which can be described by a potential function

$$V = V(q,\eta) \tag{1}$$

where q and η are vectors of dimensions n and m respectively. The set q_i $(i = 1,2,\ldots n)$ describes the behaviour of the system and may accordingly be called behaviour variables (generalized coordinates). The η_α $(\alpha = 1,2,\ldots m)$ represent m independent parameters (e.g., external loads). In structural mechanics, the total potential energy is the potential function (1). In other disciplines, one may be able to construct appropriate potentials capable of representing a given system provided it is a gradient system.

It will be assumed here that V is a real, single-valued and continuously differentiable function of q_i and η_α, at least in the region of interest, and that

$$\text{grad}_q V = 0 \tag{2}$$

describes certain equilibrium surfaces in the (q,η) space. An equilibrium state on this surface is called critical if the Hessian of the potential function at that point vanishes,

$$\det \left| \frac{\partial^2 V}{\partial q_i \partial q_j} \right| = 0 \ . \tag{3}$$

Introduce now the local coordinate system (u,λ) in the vicinity of an arbitrary equilibrium state e by setting

$$q = q^e + Au, \qquad \eta = \eta^e + \lambda \tag{4}$$

where the matrix A is composed of the orthonormalized eigenvectors associated with the Hessian matrix. Under this transformation, the potential function takes the form

$$H(u,\lambda) \equiv V(q^e + Au, \eta^e + \lambda) \tag{5}$$

where the Hessian matrix $[H_{ij}]_e$ is diagonal and the u_i are the principal coordinates.

If the point e is critical satisfying (3) then $r(r < n)$ eigenvalues of the Hessian matrix vanish simultaneously and the poine e is called an r-fold coincident critical point, simple (or distinct) points being a special case $(r = 1)$. Expanding the potential function (5) into Taylor series around an r-fold coincident point e, one obtains

$$H = \frac{1}{2!} [h_a(u_a)^2 + h_s(u_s)^2 + 2 H_{i\alpha} u_i \lambda_\alpha]$$

$$+ \frac{1}{3!} [H_{ijk} u_i u_j u_k + 3 H_{ij\alpha} u_i u_j \lambda_\alpha + \ldots]$$

$$+ \ldots \tag{6}$$

where the coefficients h_i, H_i, H_{ijk}, etc. are the derivatives of H evaluted at e (e.g. $H_{i\alpha} = \partial^2 H/\partial u_i \partial\lambda_\alpha|_e$), summation convention applies and, in view of (2), irrelevant terms are omitted. The h_i $(i = 1,2,\ldots,n)$ are the eigenvalues of the Hessian matrix, and it is assumed that $h_a = 0$ for $a = 1,2,\ldots,r$ while s ranges from r + 1 to n.

Applying $\text{grad}_u H = 0$ to (6) results in the equilibrium equations

$$H_{a\alpha}\lambda_\alpha + \frac{1}{2}(H_{aij}u_i u_j + 2H_{ai\alpha}u_i\lambda_\alpha + \ldots) + \ldots = 0 \tag{7}$$

and

$$h_{\underline{s}} u_s + H_{s\alpha}\lambda_\alpha + \frac{1}{2}(H_{sij}u_i u_j + 2H_{si\alpha}u_i\lambda_\alpha + \ldots) + \ldots = 0 \tag{8}$$

corresponding to critical (u_a) and noncritical coordinates (u_s) respectively. Here, the bar under s suspends the summation.

Evidently, the bilinar form

$$H_{a\alpha}u_a\lambda_\alpha \qquad (a = 1,2,\ldots r;\quad \alpha = 1,2,\ldots,m) \tag{9}$$

plays an important role in forming the equilibrium surface in the vicinity of the critical point e, and the critical conditions will be classified accordingly. Thus, if the rank of this bilinear form is r, the critical point will be called <u>general</u>, and if the rank is less than r it will be called <u>special</u>. The significance of the rank associated with (9) lies in the fact that the surface around the critical point is a proper (improper) one in the former (latter) case. It can be shown that, by introducing appropriate transformations of u_a and λ_α, the matrix $H_{a\alpha}$ can be canonized as follows:

$$H_{a\alpha} \sim -\,[I \quad 0] \quad \text{if the rank } k = r < m \tag{10}$$

$$H_{a\alpha} \sim - \begin{bmatrix} I \\ 0 \end{bmatrix} \qquad \text{if the rank } k = m < r \tag{11}$$

$$H_{a\alpha} \sim - \begin{bmatrix} I & 0 \\ 0 & 0 \end{bmatrix} \qquad \text{if the rank } k < \min(m,r) \tag{12}$$

where I is a k x k identity matrix, and 0 represents a null matrix.

In the first case, for example, the potential function can be written as

$$H = \frac{1}{2!}\left[h_s (u_s)^2 - 2\,\delta_{a\varepsilon}\, u_a\, \lambda_\varepsilon + 2\, H_{s\alpha}\, u_s\, \lambda_\alpha\right]$$
$$+ \frac{1}{3!}\left[H_{ijk}\, u_i\, u_j\, u_k + 2\, H_{ab\nu}\, u_a\, u_b\, \lambda_\nu\right.$$
$$\left. + 6\, H_{sa\nu}\, u_a\, u_s\, \lambda_\nu + \ldots\right] \tag{13}$$

where $\delta_{a\varepsilon}$ is the Kronecker's delta, $\varepsilon = 1,2,\ldots,r$ and $\nu = r+1,\ldots,m$. It is noted that the transformation leading to the canonical form (10) results in a distinction between the parameters λ_α. Thus, two groups of parameters, denoted by $\lambda_\varepsilon (\varepsilon = 1,2,\ldots,r)$ and $\lambda_\nu (\nu = r+1,\ldots,m)$ in (13), can be identified such that the former (latter) group is associated with a linear (quadratic) form of the critical coordinates u_a in the potential function (13), and can accordingly be called primary (secondary).

Eliminating the non-critical coordinates u_s between the equilibrium equations, it can be shown that the equilibrium surface in the vicinity of the critical point e (which is now general) in the (u_a, λ_α) space is smooth with a full complement of well-defined normals. In the remaining cases [i.e. cases (11) and (12)], some normals of the surface do not exist and the surface may consist of intersecting parts [8, 9]. In this article, attention will be focussed on general points, and the multiple-parameter perturbation technique [10] will now be employed to obtain the equations of the equilibrium surface in the vicinity of the r-fold general point e intrinsically.

1.3 Equilibrium surface

It is assumed that the equilibrium equations

$$H_i(u,\lambda) = 0, \qquad i = 1,2,\ldots,n \tag{14}$$

can be solved simultaneously to obtain the equilibrium surface in the parametric form

$$u = u(\sigma), \qquad \lambda = \lambda(\sigma) \tag{15}$$

where σ represents a set of m parameters, $\sigma = 0$ giving the critical point e. The perturbation parameters σ must be chosen such that the functions (15) are single-valued. It can be shown that, in view of (10), σ can be formed from the basic variables, $u_a (a = 1,\ldots,r)$ and $\lambda_\nu (\nu = r + 1,\ldots,m)$, and hence the functions (15) are expressed as

$$u_s = u_s(u_a,\lambda_\nu), \quad \lambda_\varepsilon = \lambda_\varepsilon(u_a,\lambda_\nu) . \tag{16}$$

Substituting these assumed solutions back into the equilibrium equations (14) yields the identities

$$H_i[u_j(u_a,\lambda_\nu), \lambda_\alpha(u_a,\lambda_\nu)] \equiv 0 \tag{17}$$

where $u_j(u_a,\lambda_\nu)$ and $\lambda_\alpha(u_a,\lambda_\nu)$ reduce to u_a and λ_ν for $j = a$ and $\alpha = \nu$ respectively.

Successive differentiations of (17) with respect to the independent variables u_a and λ_ν generate a sequence of equations. Each step of the sequence yields a set of linear equations in surface derivatives which can be solved recursively. Thus, the first and second order equations generated from (17) can be written as

$$\left.\begin{array}{l} H_{ij}\, u_{j,a} + H_{i\alpha}\, \lambda_{\alpha,a} = 0 \\ H_{ij}\, u_{j,\nu} + H_{i\alpha}\, \lambda_{\alpha,\nu} = 0 \end{array}\right\} \tag{18}$$

and

$$(H_{ijk}\, u_{k,b} + H_{ij\alpha}\, \lambda_{\alpha,b})\, u_{j,a} + H_{ij}\, u_{j,ab} +$$

$$(H_{ij\alpha}\, u_{j,b} + H_{i\alpha\beta}\, \lambda_{\alpha,b})\, \lambda_{\alpha,a} + H_{i\alpha}\, \lambda_{\alpha,ab} = 0, \tag{19a}$$

$$(H_{ijk}\, u_{k,\nu} + H_{ij\alpha}\, \lambda_{\alpha,\nu})\, u_{j,a} + H_{ij}\, u_{j,a\nu} +$$

$$(H_{ij\alpha}\, u_{j,\nu} + H_{i\alpha\beta}\, \lambda_{\beta,\nu})\, \lambda_{\alpha,\nu} + H_{i\alpha}\, \lambda_{\alpha,a\nu} = 0, \tag{19b}$$

$$(H_{ijk}\, u_{k,\mu} + H_{ij\alpha}\, \lambda_{\alpha,\mu})\, u_{j,\nu} + H_{ij}\, u_{j,\nu\mu} +$$

$$(H_{ij\alpha}\, u_{j,\mu} + H_{i\alpha\beta}\, \lambda_{\beta,\mu})\, \lambda_{\alpha,\nu} + H_{i\alpha}\, \lambda_{\alpha,\nu\mu} = 0 \tag{19c}$$

where the sets of indices $(i,j,k,\dots)$, $(a,b,\dots)$ $(\alpha,\beta,\dots)$ and $(\nu,\mu,..)$ range from 1 to n, 1 to r, 1 to m, and (r + 1) to m respectively. Subscripts on u and λ after comas indicate differentiation with respect to the corresponding independent variables.

Upon evaluation at the critical point e, the first order equations (18) yield the first order surface derivatives

$$\lambda_{\varepsilon,a} = 0, \quad \lambda_{\varepsilon,\nu} = 0, \quad u_{s,a} = 0, \quad u_{s,\nu} = -\frac{H_{s\nu}}{h_s}. \tag{20}$$

The second order equations can then be solved to yield

$$\lambda_{c,ab} = H_{cab}, \quad \lambda_{c,a\nu} = A_{ca\nu}, \quad \lambda_{c,\nu\mu} = B_{c\nu\mu} \tag{21}$$

for i = c (c = 1,2,...,r) where

$$A_{ca\nu} = H_{ca\nu} - \sum_s \frac{H_{sab}\, H_{s\nu}}{h_s},$$

$$B_{c\nu\mu} = H_{c\nu\mu} - \sum_{s} \frac{H_{sc\mu} H_{s\nu} + H_{sc\nu} H_{s\mu}}{h_s} + \sum_{s,t} \frac{H_{cst} H_{s\nu} H_{t\mu}}{h_s h_t},$$

and $i = c$ implies $c = \varepsilon$.

On the other hand, for $i = s$ $(s = r + 1, \ldots n)$ one obtains

$$u_{s,ab} = - \frac{C_{sab}}{h_s} \qquad (22)$$

where

$$C_{sab} = H_{sab} - \delta_{c\varepsilon} H_{s\varepsilon} H_{abc} .$$

The derivatives (20), (21) and (22) can now be used to construct the asymptotic equations of the equilibrium surface around the <u>general</u> critical point e as

$$\lambda_c = \frac{1}{2} H_{cab} u_a u_b + A_{ca\nu} u_a \lambda_\nu + \frac{1}{2} B_{c\nu\mu} \lambda_\nu \lambda_\mu \qquad (23)$$

and

$$u_s = - \frac{1}{h_s} (H_{s\nu} \lambda_\nu + \frac{1}{2} C_{sab} u_a u_b). \qquad (24)$$

Equation (23) represents the projection of the equilibrium surface into the (u_a, λ_α) subspace which yields significant quantitive information concerning the local behaviour of the system. For example, the second fundamental tensors associated with the system of r orthogonal normals can readily be derived from (23) which may then be used to obtain geometric invariants such as principal curvatures, lines of curvatures, asymptotic lines etc. The convexity properties of the surface established this way can later be linked to those of the stability boundary which may have important practical implications. In the following section the stability boundary will be explored on a comparative basis with the equilibrium analysis of this section. Before we proceed to do so, however, the equilibrium surface in the vicinity of a singular general point will be examined. The general point will be called singular if all the cubic coefficients H_{abc} vanish simultaneously at that point. This may happen due to certain symmetry properties of the system. Let e be singular; the derivatives $\lambda_{c,ab}$ will then vanish, and a further step (third) in the sequence of perturbations is required to determine the higher order derivatives. Following the established procedure one obtains [9]

$$\lambda_{d,abc} = H_{abcd} - 3\sum_s \frac{H_{sda} H_{sbc}}{h_s} \equiv D_{\alpha abc}$$

The first order equation of the equilibrium surface corresponding to equation (23) is then constructed as

$$\lambda_d = \frac{1}{6} D_{dabc} u_a u_b u_c + A_{ca\nu} u_a \lambda_\nu + \frac{1}{2} B_{c\nu\mu} \lambda_\nu \lambda_\mu \qquad (25)$$

1.4 Stability boundary

The critical points on the equilibrium surface, satisfying

$$\left. \begin{aligned} & H_i(u,\lambda) = 0 \\ \text{and} \quad & \Delta(u,\lambda) \equiv \det|H_{ij}(u,\lambda)| = 0 \end{aligned} \right\}$$

form the critical zone. The projection of the critical zone into the parameter space gives the critical surfaces, and the portion(s) of the critical surfaces associated with an initial loss of stability is defined as the stability boundary. In other words, the stability boundary consists of primary critical points in the parameter space whose origin is assumed to represent a stable initial state.

In order to obtain the critical zone (and hence the stability boundary) in the vicinity of the general critical point e, the criticality condition $\Delta(u,\lambda) = 0$ can be expanded into Taylor series and solved simultaneously with the equilibrium equations. Alternatively, the perturbation technique employed in section 1.3 can be adopted. In this regard, one observes that, in view of an additional equation (criticality condition), the number of independent variables is now reduced by one, and the solutions can be expressed as

$$\overset{*}{u} = \overset{*}{u}(\sigma), \quad \overset{*}{\lambda} = \overset{*}{\lambda}(\sigma) \tag{26}$$

where σ represents a set of (m - 1) parameters, and the star is used to indicate critical values. In this generality, however, the equations of the critical zone and the stability boundary can only be obtained in a parametric form. In order to form an idea, consider the case in which the general critical point is simple such that $h_a = 0$ consists of $h_1 = 0$, and u_1 is the critical coordinate. Then, the equilibrium equation (23) takes the form

$$\lambda_1 = \frac{1}{2} H_{111} u_1^2 + A_{11\nu} u_1 \lambda_\nu + \frac{1}{2} B_{1\nu\mu} \lambda_\nu \lambda_\mu , \tag{27}$$

and the corresponding critical zone in this case can be obtained directly by differentiating (27) with respect to u_1 as

$$\overset{*}{u}_1 = - \frac{1}{H_{111}} A_{11\nu} \overset{*}{\lambda}_\nu \tag{28}$$

which together with (27) yields

$$\overset{*}{\lambda}_1 = \frac{1}{2 H_{111}} (A_{11\nu} A_{11\mu} - H_{111} B_{1\nu\mu}) \overset{*}{\lambda}_\nu \overset{*}{\lambda}_\mu . \tag{29}$$

Equation (29) represents the stability boundary if the critical point is primary so that all $h_s > 0$. Equations (28) and (29) can also be obtained by choosing the functions (26) as

$$\overset{*}{u}_s = \overset{*}{u}_s(\overset{*}{\lambda}_\nu), \qquad \overset{*}{\lambda}_1 = \overset{*}{\lambda}_1(\overset{*}{\lambda}_\nu),$$

and perturbing the identities

$$\begin{aligned} &H_i[\overset{*}{u}_j(\overset{*}{\lambda}_\nu), \overset{*}{\lambda}_1(\overset{*}{\lambda}_\nu)] \equiv 0 \\ &\Delta[\overset{*}{u}_i(\overset{*}{\lambda}_\nu), \overset{*}{\lambda}_1(\overset{*}{\lambda}_\nu)] \equiv 0 \end{aligned} \tag{30}$$

to generate necessary surface derivatives as before (see reference [5]).

If the system has only one degree of freedom and the potential energy is linear in parameters (which is usually the case), then (27) takes the form

$$\lambda_1 = \frac{1}{2} H_{111} u_1^2 + H_{11\nu} u_1 \lambda_\nu \tag{31}$$

and (29) reduces to

$$\overset{*}{\lambda}_1 = \frac{H_{11\nu} H_{11\mu}}{2 H_{111}} \overset{*}{\lambda}_\nu \overset{*}{\lambda}_\mu \equiv \frac{(H_{11\nu} \overset{*}{\lambda}_\nu)^2}{2 H_{111}}. \tag{32}$$

Observing that the curvatures of the stability boundary,

$$\overset{*}{\lambda}_{1,\nu\nu} = \frac{(H_{11\nu})^2}{H_{111}},$$

and the curvature of the equilibrium surface (31)

$$\lambda_{1,11} = - H_{111}, \tag{33}$$

will always have opposite signs, the following theorem can be stated:

Theorem 1.

The stability boundary of a one-degree of freedom system associated with general critical points cannot have convexity towards the region of instability.

Several other general properties can be established (see reference [5]). Here, it is finally observed that in the case of a simple singular point, the equilibrium surface (25) takes the form

$$\lambda_1 = \frac{1}{6} D_{1111} u_1^3 + A_{11\nu} u_1 \lambda_\nu + \frac{1}{2} B_{1\nu\mu} \lambda_\nu \lambda_\mu, \tag{34}$$

and the corresponding critical zone is obtained by differentiating (34) with respect to u_1 as

$$\frac{1}{2} D_{1111} \overset{*}{u}_1^2 = - A_{11\nu} \overset{*}{\lambda}_\nu, \tag{35}$$

which together with (34) leads to

$$A_{11\nu} \overset{*}{\lambda}_\nu = - \frac{1}{2} D_{1111}^{1/3} (3 \overset{*}{\lambda}_1)^{2/3}. \tag{36}$$

Equation (36) defines the stability boundary and represents a generalization of the familiar two-thirds power law associated with imperfect systems [1].

1.5 Geometry of the equilibrium surface and connection with the catastrophe theory

Consider first the simplest case which is associated with a distinct general point. The corresponding equilibrium surface is described by (27). This surface was shown [7] to be equivalent to Fold Catastrophe which can essentially be represented by one parameter only if a qualitative description is sought as in the case of catastrophe theory. In a quantitative analysis, however, the connexity properties and curvatures have significant practical implications, and it is noted that the equilibrium surface in the form of (27) contains such information. Thus, the surface is synclastic, anticlastic or parabolic according to whether the matrix

$$[A_{11\nu} A_{11\mu} - H_{111} B_{1\nu\mu}]$$

is negative definite, positive definite or null respectively. It is interesting that a general critical point appears as a limit point on λ_1 - u_1 plane and

and as a point of bifurcation on a λ_ν - u_1 plane which may lead to confusion with regard to the nature of instability (Figure 1). This was one of the underlying motivations of proposing a reclassification of critical conditions and introducing the concept of a general critical point in the first place [4]. According to this classification, instabilities are associated with certain definite forms of surfaces (such as the one in Figure 1), and the descriptions are not limited to the inadequate framework of one-parameter phenomena.

The next distinct type of surface is associated with a simple singular point. The equation of this surface is given by (34) and it was shown [7] that this surface corresponds to Riemann-Hugoniot Catastrophe. Due to its significance, the necessary transformations will be explained here as well. It is first noted that the behaviour of the system remains qualitatively similar for all one-parameter control paths $\lambda_\nu = \lambda_\nu(\xi)$ where ξ is a variable parameter. Without loss of generality, one can then consider λ_2 as a representative parameter, thus reducing (34) to the two-parameter equation

$$\lambda_1 = \frac{1}{6} D_{1111} u_1^3 + A_{112} u_1 \lambda_2 + \frac{1}{2} B_{122} \lambda_2^2. \tag{37}$$

Setting $\lambda_1 = 0$ and differentiating (37) with respect to u_1 twice yields upon evaluation at the singular point e

$$\left.\frac{d\lambda_2}{du_1}\right|_e = 0 \qquad \text{or} \qquad \left.\frac{d\lambda_2}{du_1}\right|_e = -2\,\frac{A_{112}}{B_{122}}$$

which indicate a symmetric point of bifurcation in λ_2 - u_1 plane. Based on this property, one can introduce the sliding and rotating coordinates

$$u = u^e(\lambda_2) + T(\lambda_2)\bar{u} \tag{38}$$

where $u^e(\lambda_2)$ is the single-valued fundamental path through e and the transformation matrix $T(\lambda_2)$ keeps the diagonal form of the potential energy diagonalized along this path. If (38) is introduced into the original function $H(u,\lambda)$, a new function

$$S(\bar{u}_1,\lambda_1,\lambda_2) \equiv H[u_i^e(\lambda_2) + t_{ij}(\lambda_2)\bar{u}_j,\lambda_1,\lambda_2] \tag{39}$$

is obtained with the properties

$$\left.\begin{aligned} &S_i(0,\lambda_2,0) = S_{i2}(0,\lambda_2,0) = S_{i22}(0,\lambda_2,0) = \ldots = 0 \\ &S_{ij}(0,\lambda_2,0) = S_{ij2}(0,\lambda_2,0) = \ldots = 0 \text{ for } i \neq j \end{aligned}\right\} \tag{40}$$

Applying the multiple-parameter perturbation technique to the new function S and using the properties (40), the first order equation of the equilibrium surface can be constructed as

$$\lambda_1 = \frac{1}{6}\bar{S}_{1111}\,\bar{u}_1^3 + S_{112}\,\bar{u}_1\,\lambda_2 \tag{41}$$

where

$$\bar{S}_{1111} = S_{1111} - 3\sum_s \frac{(S_{s11})^2}{S_s}$$

Equation (41) was obtained in reference [7] in a different way, and is recognized immediately as the Riemann-Hugoniot catastrophe. The stability boundary (bifurcation set) can readily be determined as

$$\overset{*}{\lambda}_2 = \frac{1}{2}\,\frac{(\bar{S}_{1111})^{1/3}}{S_{112}}\,(3\,\overset{*}{\lambda}_1)^{2/3} \tag{42}$$

which describes a cusp in parameter (control) space. The equilibrium surface (41) and the stability boundary (42) are shown in Figure 2. Obviously, the point e appears as a singular point in the overall picture, resulting in an abrupt change in the convexity of the boundary. The parameters λ_1 and λ_2 are identified as the shock and normal parameters of the catastrophe theory respectively, and the importance of the transformation (10) in identifying the significant parameters is observed.

It is noted that the general point and the singular general point are, in fact, associated with a surface of order two and three respectively. If further energy coefficients vanish at the critical point, higher order perturbations yield higher order surfaces. If, for example $\bar{S}_{1111} = 0$ in (41), one proceeds to construct a surface of order four, identifying the general point of order four which has been shown [7] to correspond to swallow tail of catastrophe theory. In fact, following the pattern, one can continue to generate higher order surfaces (see reference [7]).

So far, attention has been focussed on distinct (r = 1) critical points. If the general point is associated with a 2-fold eigenvalue (r = 2), then the equilibrium surface (23) is linked to elliptic or hyperbolic umbilic of catastrophe theory depending mainly on the properties of the cubic form of the potential function (in u_1 and u_2). Noting again that the behaviour of the system remains qualitatively similar for arbitrary rays $\lambda_\nu = \ell_\nu \xi$ (where ℓ_ν are constants), the equation (23) takes the form

$$\lambda_1 = \frac{1}{2} H_{111} u_1^2 + H_{122} u_1 u_2 + \frac{1}{2} H_{122} u_2^2 + A_{11} u_1 \xi + A_{12} u_2 \xi + \frac{1}{2} B_1 \xi^2 \tag{43}$$

$$\lambda_2 = \frac{1}{2} H_{211} u_1^2 + H_{212} u_1 u_2 + \frac{1}{2} H_{222} u_2^2 + A_{21} u_1 \xi + A_{22} u_2 \xi + \frac{1}{2} B_2 \xi^2 \tag{44}$$

where $A_{11} = A_{11\nu} \ell_\nu$, $A_{12} = A_{21} = A_{12\nu} \ell_\nu$, $A_{22} = A_{22\nu} \ell_\nu$,

$B_1 = B_{1\nu\mu} \ell_\nu \ell_\mu$, $B_2 = B_{2\nu\mu} \ell_\nu \ell_\mu$

These equations are assumed to be derived from the potential

$$P = \frac{1}{6} H_{111} u_1^3 + \frac{1}{2} H_{122} u_1^2 u_2 + \frac{1}{2} H_{122} u_1 u_2^2 + \frac{1}{6} H_{222} u_2^3$$

$$+ \frac{1}{2} \xi (A_{11} u_1^2 + 2 A_{12} u_1 u_2 + A_{22} u_2^2) - u_1 \phi_1 - u_2 \phi_2 \qquad (45)$$

where $\phi_1 = \lambda_1 - \frac{1}{2} B_1 \xi^2$ and $\phi_2 = \lambda_2 - \frac{1}{2} B_2 \xi^2$.

The cubic form of (45) in u_1 and u_2 can be expressed as the product of three linear forms. If these forms are linearly independent and all real (two of them complex conjugate) the general point corresponds to elliptic umbilic (hyperbolic umbilic) provided the quadratic form associated with ξ is positive definite (indefinite). The analysis is given in reference [11] and will not be repeated here. Similarly, if only two of the linear forms are linearly independent the behaviour is normally associated with parabolic umbilic.

In numerical applications, another way of ascertaining the character of an equilibrium surface is to determine the principal curvatures of the surface at the critical point and compare with those of umbilics. In this regard, it can readily be established [9] that the principal curvatures of the hyperbolic umbilic are given by (-1,1,6) and (-1,1,6), while those of elliptic umbilic are $(6,-3\pm\sqrt{13})$ and $(0,\pm 2\sqrt{10})$.

1.6 Examples

(i) An imperfect model

References [5, 7] contain several applications and illustrative examples of the theory discussed in the preceding sections. The two-degree-of-freedom imperfect model shown in Figure 3 will be considered here as an example of the singular case (Riemann-Hugoniot Catastrophe). The model is under the axial force and small horizontal force ε, and consists of two pin-jointed rigid members. Each joint has a linear torsional spring with $k_1 = 2k$ and $k_2 = k$. The total potential energy may be expressed [5] as

$$V = k\, q_1^2 + \frac{1}{2} k(q_2 - q_1)^2$$

$$- \Lambda\, \ell[\frac{1}{2} (q_1^2 + q_2^2) - \frac{1}{24} (q_1^4 + q_2^4) + \ldots]$$

$$- \varepsilon\, \ell(q_1 + \frac{1}{3!} q_1^3 + \ldots). \tag{46}$$

Although the model has two behaviour variables (q_1 and q_2), the system is not of dimension two in terms of catastrophe theory, and none of the universal unfoldings is directly applicable to the potential (46). In order to elucidate the behaviour of the system, the systematic stability analysis discussed in the theory must be applied. Thus, the potential energy is first transformed into appropriate canonical form by first observing that the equilibrium equations obtained from (46) yield $q_i = 0$ as an equilibrium path for $\varepsilon = 0$, and then introducing the orthogonal transformation

$$q_i = t_{ij}\, u_j \text{ where } t_{ij} = \begin{bmatrix} 0.3827 & 0.9239 \\ 0.9239 & -0.3827 \end{bmatrix}$$

to have the transformed function (39)

$$S(u_i, \Lambda, \varepsilon) = \frac{1}{2} k(0.59\, u_1^2 + 3.41\, u_2^2) - \Lambda\, \ell\Big\{\frac{1}{2} (u_1^2 + u_2^2)$$

$$- \frac{1}{24} [(0.38\, u_1 + 0.92\, u_2)^4 + (0.92\, u_1 - 0.38\, u_2)^4]\Big\}$$

$$- \varepsilon\, \ell(0.38\, u_1 + 0.92\, u_2 + \frac{1}{3!} \ldots). \tag{47}$$

One can then evaluate the derivatives

$$S_{11}(0,\Lambda,0) = 0.59\, k - \ell\, \Lambda$$

$$S_{11}(0,\Lambda,0) = 3.4\, k - \ell\, \Lambda$$

which yield the primary critical load

$$\Lambda_{cr} = 0.59\ k/\ell \tag{48}$$

and indicate the critical coordinate u_1.

Using the theory one can readily obtain the equation of the equilibrium surface in the vicinity of the critical point (48) as

$$0.38\ \ell\ \varepsilon + \frac{0.44}{3!}\ k\ u_1^3 - \ell\ u_1\ \lambda = 0 \tag{49}$$

which is identical to (41) with $\lambda_1 = \varepsilon$ and $\lambda_2 = \lambda = \Lambda - \Lambda_{cr}$, indicating a singular point or Riemann-Hugoniot catastrophe. It is emphasized once more that despite the fact that the original system has two behaviour variables, the equilibrium surface (49) is in terms of u_1, the critical coordinate only. The surface is similar to that shown in Figure 2, and the stability boundary (42) is given as

$$\lambda = \frac{1}{2\ell}\ (0.44\ k)^{1/3}\ (3 \times 0.38\ \ell\ \varepsilon)^{2/3} \tag{50}$$

(ii) Phase transition

In order to indicate the wide range of applicability of the multiple-parameter stability theory, the phenomena of liquid/gas/liquid transitions in Thermodynamics will now be discussed. Consider the Van der Waals equation

$$\left(P + \frac{a}{V^2}\right)(V - b) = cT \tag{51}$$

where P,V,T denote pressure, volume and temperatures respectively and a,b,c are constants. Treating (51) as an equilibrium equation relating the control parameters P and T to behaviour variable V, one can proceed with higher order derivatives with respect to the latter to find the critical points and identify the system according to the general theory. To this end, write (51) in the form

$$PV^3 - (bP + cT)V^2 + aV - ab = 0 \tag{52}$$

and differentiate (52) with respect to V twice to obtain

$$3PV^2 - 2(bP + cT)V + a = 0 \tag{53}$$

and

$$6PV - 2(bP + cT) = 0 . \tag{54}$$

The simultaneous solution of these three equations yields the critical point e in the (P,T,V) space as

$$V_{cr} = 3b, \quad P_{cr} = \frac{a}{27b^2}, \quad T_{cr} = \frac{8a}{27bc} \tag{55}$$

which is singular since the cubic term vanishes as in the theory.

The coordinates can be shifted to the singular point (55) and rotated according to the transformation (not orthogonal)

$$\begin{bmatrix} \frac{P}{P_{cr}} \\ \\ \frac{T}{T_{cr}} \end{bmatrix} = \begin{bmatrix} 1 \\ \\ 1 \end{bmatrix} + \frac{1}{3} \begin{bmatrix} -1 & 1 \\ \\ \frac{1}{8} & \frac{1}{4} \end{bmatrix} \begin{bmatrix} \lambda_1 \\ \\ \lambda_2 \end{bmatrix} \tag{56}$$

$$u = \frac{V_{cr}}{V} - 1$$

which transforms the Van der Waals equation in the vicinity of the singular point into

$$\lambda_1 + 3\,u^3 + u\,\lambda_2 = 0 . \tag{57}$$

Here the density u is used as the behaviour variable which is more appropriate [12]. The surface described by (57) is shown in Figure 4. Phase transition from liquid to gas and/or from gas to liquid occurs on the critical zone given by (57) and

$$9\,u^2 + \lambda_2 = 0 , \tag{58}$$

The stability boundary in the parameter space can be obtained by eliminating u between (57) and (58). Physically, the jumps occurring on the critical line (Figure 4) represent boiling and condensation. In changing from liquid to gas, for example, the same state can also be reached by a smooth path rather than jumps if the parameters (pressure and temperature) are controlled such that the critical zone is avoided. The interior sheet of the surface represents unstable states referred to as "gaquid".

Several other examples can be found in reference [7].

2. Stability of Autonomous Systems

2.1 Definitions

Consider an autonomous holonomic dynamical system described by the generalized coordinates $q_i (i = 1,2,\ldots,n)$ and independent parameters $\eta_\alpha (\alpha = 1,2,\ldots,m)$. The free motion of this system is governed by n ordinary differential equations of second order (Lagrange's equations). Alternatively, the motion can be described in the state space by a set of first order differential equations. For a given set of η_α, the latter equations can be expressed in the vector form

$$\dot{x} = f(x) \tag{59}$$

where x is the state vector of order 2n, having the generalized coordinates q_i and their time derivatives $\dot{q}_i$ as its elements and $\dot{x} = dx/dt$. An equilibrium state x^e satisfies $f(x^e) = 0$ for all t.

Without loss of generality, suppose $x = 0$ is an equilibrium state and that its stability is to be investigated. According to Liapunov, the state $x = 0$ is stable if it is possible to find a positive $\delta(\varepsilon)$ for any sufficiently small $\varepsilon > 0$ such that following the initial disturbance

$$||x^\circ|| < \delta \quad \text{where } x^\circ = x(0)$$

the inequality

$$||x(t)|| < \varepsilon$$

is satisfied for all $t > 0$. Otherwise the state $x = 0$ is unstable.

The state x = 0 is said to be asymptotically stable if it is stable and, in addition, one has

$$\lim_{t\to\infty} ||x(t)|| = 0.$$

It is often desired to obtain information concerning the stability oroperties of the system without actually solving the differential equations. For this purpose Liapunov formulated certain theorems. Here two of these theorems are given:

(a) Stability theorem. The Null solution of the system (59) is stable if in some region R about the origin there exists a Liapunov function V(x).

(b) Asymptotic stability theorem. The Null solution is asymptotically stable if the derivative $\dot{V}(x)$ of the Liapunov function V(x) is negative definite.

A real scalar function of the vector x which is continuous with its first partial derivatives in a certain region R containing origin, is said to be a Liapunov function if it is positive definite in R while its time derivative is negative semi-definite in R.

The proofs of these theorems can be found in several books [13, 14]. For other definitions and related discussion see, for example, Leipholz's book [15] and his article in this volume.

It is inferred from (59) that the definition and criteria of stability given above are not restricted to linear systems although they are related to local properties. In this article, however, attention will be focussed on linear systems in which case (59) takes the form

$$\dot{x} = A\,x \tag{60}$$

where A is a square matrix of order 2n.

It can be shown [14] that a necessary and sufficient condition for the null solution x = 0 of (60) to be asymptotically stable is that all the eigenvalues of A have negative real parts. If at least one of the eigenvalues has positive real part, the solution is unbounded and hence unstable. If the matrix A has no eigenvalues with positive real part but some eigenvalues have zero real part, then asymptotic stability is ruled out but not stability. The multiplicity of eigenvalues and linear dependence of the corresponding eigenvectors play a significant role on the stability of the system in this case. If all the eigenvectors are linearly independent the solution remains bounded regardless multiplicity of eigenvalues. On the other hand, a reduction in the dimension of the eigenvector space is always associated with a corresponding multiplicity of eigenvalues. It can be shown [14] that when the index of all eigenvalues is equal to their multiplicity, the eigenvectors of A are all linearly independent and the matrix A is of simple structure. These arguments apply to the linear system (60) when it is considered on its own. In practice, however, the linear system (60) is often a linearized model of a nonlinear system whose stability behaviour, in the case when some eigenvalues have zero real part (while the remaining eigenvalues have negative real parts), may not coincide with those of the linearized model. It may, therefore, be appropriate to call such cases critical, implying that higher order terms may be necessary to decide on stability. On the other hand, the term critical may not be appropriate in many important situations falling within this case, and this point will further be discussed with regard to second order differential equations in the sequel.

In practice, one often formulates a problem through Lagrange's equations which result in the linear second order equations

$$M\ddot{q} + B\dot{q} + Cq = 0 \tag{61}$$

where the n x 1 vector q represents the generalized coordinates, and M,B,C are constant n-square matrices. Normally, M is a symmetric positive definite matrix and the systems can be classified according to the properties of the matrices B and C.

First note the connection between (61) and the first order differential equation (60). Evidently, if A is defined by

$$A = \begin{bmatrix} -M^{-1}B & -M^{-1}C \\ I & 0 \end{bmatrix} \tag{62}$$

and the state vector x by

$$x = \begin{bmatrix} \dot{q} \\ q \end{bmatrix} \tag{63}$$

equations (60) and (61) become equivalent.

Introducing the solution

$$q = u\, e^{\lambda t}$$

into (61) yields*

$$(M\lambda^2 + B\lambda + C)u = 0 \tag{64}$$

which can be transformed into the eigenvalue problem

$$(A - \lambda I)z = 0 \tag{65}$$

*Note that the notation in chapter 1 and chapter 2 of this article is defined separately and λ of part 1, for example, has nothing to do with λ here.

where A is given by (62) and

$$z = \begin{bmatrix} \lambda u \\ u \end{bmatrix} \tag{66}$$

Nontrivial solutions of (64) exist only if the matrix $(M\lambda^2 + B\lambda + C)$ is singular, that is, if λ is a root of the characteristic equation

$$\Delta(\lambda) = \det|M\lambda^2 + B\lambda + C| = 0. \tag{67}$$

If the matrices M, B, and C are real, distinct roots occur in complex conjugate pairs

$$\lambda = \gamma \pm i\,\omega$$

and the corresponding eigenvectors are also complex conjugate.

On the basis of the roots

$$\lambda_j = \gamma_j + i\,\omega_j \qquad j = 1,2,\dots,2n$$

the following three cases are identified with regard to the stability of the equilibrium state $q = \dot{q} = 0$:

(a) All $\gamma_j < 0$: asymptotic stability

(b) At least one $\gamma_k > 0$: instability

(c) If some $\gamma_k = 0$ while the remaining γ's are negative, the state is in general critical, particularly if the system is a linearized model of a nonlinear system. Asymptotic stability is ruled out but not stability. The following special cases are of particular interest:

(i) If all $\gamma_j = 0$ and ω is not repeated or more generally, if the multiplicity of a repeated root is equal to the degeneracy of the λ-matrix $(M\lambda^2 + B\lambda + C)$, the state is stable and this case cannot be referred to as critical [14]. If at least a pair of ω's vanishes as well (assuming complex conjugate roots), the linear system is unstable, but if it is to be considered as a linearized model of a nonlinear system this situation should be treated as critical, a decision on stability requiring higher order terms (e.g. systems in chapter 1).

(ii) If all $\gamma_j = 0$ and at least a pair of ω's coincide such that there is a reduction in the dimension of the eigenvector space, then the linear system is unstable.

In the following analysis two types of instability will be identified:

1. Divergence: this instability is defined by

   some $\gamma_j > 0, \quad \omega_j = 0$

2. Flutter: Flutter occurs if

   some $\gamma_j > 0, \quad \omega_j \neq 0$.

Similarly, the critical state $\gamma_j = 0$ is also called divergence or flutter according to above definitions (e.g. cases (i) and (ii) respectively). In a linear system divergence is associated with a branching of solution shile flutter is a dynamic loss of stability characterized by oscillations with increasing amplitude.

2.2 Classification of systems

The linear autonomous systems under consideration can be classified on the basis of the properties of the matrices in (64) for a more detailed analysis. Expressing (64) as

$$[M\lambda^2 + (D + G)\lambda + (K + S)]u = 0, \tag{68}$$

where $D = D' \geq 0$ (symmetric positive semi-definite)

$G = -G'$, $K = K'$ and $S = -S'$,

the following four classes of systems will be considered here:

1. Conservative systems: described by (68) with $D = G = S = 0$.
2. Pseudo-conservative systems: described by (68) with $D = G = 0$ and

$M^{-1}(K + S)$ symmetrizable.

3. Gyroscopic systems: described by (68) with D = S = 0.

4. Circulatory systems: described by (68) with D = G = 0.

In each case the effect of damping (represented by D) will be discussed briefly.

Before exploring each class individually a general stability criterion can be established. Thus, pre-multiplying (68) by the complex conjugate vector $\bar{u}'$ yields

$$\lambda^2 + \lambda(d + ig) + k + is = 0 \tag{69}$$

where $\langle\bar{u},Mu\rangle = 1$, $\quad d = \langle\bar{u},Du\rangle \geq 0$

$k = \langle\bar{u},Ku\rangle$, $\quad ig = \langle\bar{u},Gu\rangle$, $\quad is = \langle\bar{u},Su\rangle$

and $\langle\rangle$ denotes an inner product. Clearly, k, g, d and s are all real. Introducing

$$\lambda = \gamma + i\omega$$

into (69) yields

$$\begin{aligned} &\gamma^2 - \omega^2 + \gamma d - \omega g + k = 0, \\ &2\gamma\omega + \gamma g + \omega d + s = 0, \end{aligned} \tag{70}$$

Eliminating ω between these equations one obtains

$$\gamma^4 + a_1\gamma^3 + a_2\gamma^2 + a_3\gamma + a_4 = 0 \tag{71}$$

where

$$a_1 = 2d, \qquad a_2 = \frac{1}{4}(5d^2 + g^2 + 4k)$$

$$a_3 = \frac{d}{4}(d^2 + g^2 + 4k), \qquad a_4 = \frac{1}{4}(kd^2 + sdg - s^2)\,.$$

According to Routh-Hurwitz criterion one requires that

$$a_i > 0 \qquad i = 1,2,3,4 \tag{72}$$

and $\quad (a_1 a_2 - a_3)a_3 - a_1^2 a_4 > 0$

for stability.

It then follows that the system is asymptotically stable if and only if

$$d > 0 \text{ and } kd^2 + sdg - s^2 > 0 \tag{73}$$

for all eigenvectors u normalized with respect to M. For undamped systems (D = 0), however, modified stability conditions have to be derived since asymptotic stability is ruled out by (73) in that case. It is not difficult to verify that with D = 0, the stability is ensured if and only if

$$g^2 + 4k > 0 \text{ and } s = 0 \ (k \neq 0) \tag{74}$$

for all eigenvectors normalized with respect to M. Similarly, if $d > 0$

$$\text{the equilibrium is } \begin{cases} \text{critical} \\ \text{unstable} \end{cases} \text{ if and only if } \begin{matrix} kd^2 + sdg - s^2 = 0 \\ kd^2 + sdg - s^2 < 0 \end{matrix} \tag{75}$$

for some mode(s). Note that the state is called critical without requiring it to be primary. For d = 0, the definitions of criticality and instability follow from (74). In this case, k = 0 is associated with a critical divergent point and is, therefore, excluded in the definition of stability in (74).

2.3 Conservative systems

The motion of a conservative system in the vicinity of the equilibrium state $q = \dot{q} = 0$ is governed by

$$M\ddot{q} + Kq = 0$$

or, upon assuming $q = u\, e^{i\omega t}$, by

$$(-M\omega^2 + K)u = 0 \tag{76}$$

which is stable if and only if $k > 0$ according to (74). Since $M > 0$ and $K = K'$, the eigenvalues ω^2 are always real and flutter instability is ruled out. The matrix $M^{-1}K$ is symmetrizable [14], and $\omega^2 > 0$ as long as $K > 0$. If some $\omega^2 < 0$ the state is unstable while $\omega^2 = 0$ corresponds to a critical state.

If the system is under the influence of several independent parameters $\eta_\alpha(\alpha = 1,2,\ldots m)$ the matrix K is in the form

$$K = U - \eta_\alpha E^\alpha$$

where $U = U' > 0$, $E = E'$ for all α, and summation convention is adopted. The characteristic equation

$$\Delta(\omega^2,\eta_\alpha) \equiv |-\omega^2 M + U - \eta_\alpha E^\alpha| = 0 \tag{77}$$

defines certain m dimensional characteristic surfaces in the m + 1 dimensional $\omega^2 - \eta_\alpha$ space, and the part of the surfaces which consists of points closest to the origin form the fundamental characteristic surface (f.c.s.). The stability boundary is associated with the (f.c.s.) and can be obtained as the intersection of the surface with the plane $\omega^2 = 0$. The intersections of the characteristic susrfaces with $\omega^2 = 0$ will, in general, be called divergence boundary which contains stability boundary.

It will now be shown that the (f.c.s.) is a strictly convex surface which cannot have convexity toward the fundamental region containing the origin of $\omega^2 - \eta_\alpha$ space. This can be proved in several ways [14].

Consider the eigenvalue problem (76); for fixed values of η_α, the frequencies of free vibrations can be expressed as

$$\omega^2 = \frac{\langle u,(U - \eta_\alpha E^\alpha)u\rangle}{\langle u,Mu\rangle} \tag{78}$$

Setting $\eta_\alpha = 0$, one obtains the natural frequencies

$$\omega^2 = \frac{\langle u,Uu\rangle}{\langle u,Mu\rangle}$$

which are obviously positive.

The eigenvalues ω_i^2 $(i = 1,2,\ldots,n)$ of the stationary values of the Rayleigh Quotient

$$R(x) = \frac{\langle x,(U - \eta_\alpha E\alpha)x\rangle}{\langle x,Mx\rangle} \tag{79}$$

where x is an arbitrary real n-vector. Also, if the eigenvalues are in the ascending order

$$\omega_1^2 \leq \omega_2^2 \leq \ldots \leq \omega_n^2, \tag{80}$$

the extremum properties of Rayleigh Quotient yield

$$\omega_1^2 \leq R(x) \leq \omega_n^2 \ . \tag{81}$$

Consider now an arbitrary point C $(\omega_0^2,\eta_\alpha^0)$ on the (f.c.s.) and denote the corresponding eigenvector by x (Figure 5). It can be shown that the equation

$$\langle x,(-\omega^2 M + U - \eta_\alpha E^\alpha)x\rangle = 0 \tag{82}$$

describes a plane tangent to the fundamental surface at the point C. Equation (82) does indeed define a plane in $\omega^2 - \eta_\alpha$ space, and $(\omega_0^2,\eta_\alpha^0)$ satisfies this equation by virtue of (76), x being the corresponding eigenvector. To show that (82) is tangential to the (f.c.s.), differentiate (76) with respect to $\eta_\alpha (\alpha = 1,2,\ldots,m)$ to obtain

$$\left(-\frac{\partial\omega^2}{\partial\eta_\alpha} M - E^\alpha\right)u + (-\omega^2 M + U - \eta_\alpha E^\alpha)\frac{\partial u}{\partial\eta_\alpha} = 0$$

which upon pre-multiplying by u and evaluating at C results in

$$\langle x,\left(-\frac{\partial\omega^2}{\partial\eta_\alpha} M - E^\alpha\right)x\rangle = 0$$

giving the slopes of the (f.c.s.) as

$$\left.\frac{\partial^2\omega^2}{\partial\eta_\alpha}\right|_C = -\frac{\langle x,E^\alpha x\rangle}{\langle x,Mx\rangle} \qquad \alpha = 1,2,\ldots,m\ . \tag{83}$$

It is immediately observed that the plane (82) has the same slopes and it is, therefore, tangential to the (f.c.s.) at C.

Consider next the ray

$$\left.\begin{aligned}\eta_\alpha &= \ell_\alpha\,\xi\\ \omega^2 &= \ell\,\xi\end{aligned}\right\} \tag{84}$$

passing through the origin of the $\omega^2 - \eta_\alpha$ space and intersecting the tangent plane as well as the characteristic surfaces. Here ℓ_k and ℓ are constants and ξ measures the distance from the origin. The intersection with the tangent plane (82) yields

$$\xi_t = \frac{\langle x,Ux\rangle}{\langle x,(M_1 + E)x\rangle} \quad \text{or} \quad \frac{1}{\xi_t} = \frac{\langle x,(M_1 + E)x\rangle}{\langle x,Ux\rangle} \tag{85}$$

where $M_1 = \ell M$ and $E = \ell_\alpha\, E^\alpha$.

On the other hand, the intersections with the characteristic surfaces result in

$$\frac{1}{\xi_i} = \frac{\langle u^i,(M_1 + E)u^i\rangle}{\langle u^i,Uu^i\rangle} \qquad i = 1,2,\ldots,n\ . \tag{86}$$

Assuming that

$$\frac{1}{\xi_1} \geq \frac{1}{\xi_2} \geq \cdots \geq \frac{1}{\xi_n}\ , \tag{87}$$

The extremum properties of Rayleigh Quotient yield

$$\frac{1}{\xi_1} \geq \frac{1}{\xi_t} \quad \text{or} \quad \xi_1 \leq \xi_t\ . \tag{88}$$

Recalling that C was an arbitrarily chosen point on the (f.c.s.), the following theorem can be formulated [14]:

Theorem 2. The fundamental characteristic surface cannot have convexity toward the fundamental region.

An immediate corollary of this theorem is concerned with the convexity of the stability boundary which is a part of the (f.c.s):

Corollary 2.1. The stability boundary cannot have convexity toward the fundamental region.

Other related theorems, discussion, extension and references can be found in reference [14].

Effect of damping

If damping is added to the conservative system considered in this section, one obtains the differential equations of motion

$$M\ddot{q} + D\dot{q} + Kq = 0 \tag{89}$$

where $D \geq 0$.

If $D > 0$ and $K > 0$, the system is asymptotically stable according to the criterion (73) since $d > 0$ and $k > 0$ for all eigenvectors. Furthermore, if the undamped system is unstable so is the damped system. Is it, however, necessary that $D > 0$ for asymptotic stability? What happens if $D \geq 0$? It can be shown that positive definiteness of D is sufficient but not necessary for asymptotic stability, and the eigenvalues can all have negative real parts even when D is non-negative definite.

Considering, the eigenvalue problem

$$(M\lambda^2 + D\lambda + K)u = 0, \tag{90}$$

suppose for some eigenvalue λ, $d = 0$ and λ is, therefore, imaginary. Since D is assumed to be positive semi-definite $d = 0$ implies

$$Du = 0 \tag{91}$$

Introducing (91) into (90), one observes that u is then an eigenvector of the conservative system $(M\lambda^2 + K)u = 0$ as well. Conversely, if u is an eigenvector of the above conservative system and it satisfies $Du = 0$, then u is also an eigenvector of (90). Thus, the necessary and sufficient condition that the system (90) be asymptotically stable is that none of the eigenvectors of the corresponding conservative system lies in the null space of D [14]. Systems satisfying these conditions are said to be pervasively damped.

Since both M and K are symmetric and $M > 0$, these two matrices can be diagonalized simultaneously such that the differential equation of the undamped system takes the form

$$\ddot{p} + w^2 p = 0, \quad \text{where } w^2 = \text{diag}[\omega_1^2, \omega_2^2, \ldots, \omega_n^2]$$

and the motion associated with each principal coordinate p_j is independent of the remaining coordinates. It may be interesting to know under what conditions such a complete uncoupling occurs in the damped system (89). It is known (see for example the first chapter in [14]) that two symmetric matrices can be reduced to diagonal forms by means of an orthogonal transformation if and only if they commute and that the respective eigenvalues of the matrices form the resulting diagonal matrices. It then follows, upon writing (89) as

$$\ddot{q} + M^{-1}D\dot{q} + M^{-1}Kq = 0,$$

that (89) can be uncoupled completely if and only if the matrices $M^{-1}D$ and $M^{-1}K$ commute. That is if and only if

$$M^{-1}DM^{-1}K = M^{-1}KM^{-1}D$$

or equivalently

$$DM^{-1}K = KM^{-1}D.$$

If this condition is satisfied, then one can find an orthogonal transformation $q = Tp$ $(T' = T^{-1})$ which results in

$$\ddot{p} + \Lambda\dot{p} + w^2 p = 0$$

where Λ is diagonal and consists of the eigenvalues of $M^{-1}D$. The following theorem has been proved:

Theorem 3. A damped conservative system can have uncoupled modes of vibration if and only if $DM^{-1}K = KM^{-1}D$.

2.4 Pseudo-conservative systems

The systems described by

$$M\ddot{q} + (U - \eta_\alpha E^\alpha)q = 0, \tag{92}$$

where at least some of the matrices E^α are asymmetric, will be defined as pseudo-conservative [14] if the matrix $M^{-1}(U - \eta_\alpha E^\alpha)$ is symmetrizable for all values of η_α. Equation (92) can also be written as

$$M\ddot{q} + (K + S)q = 0 \tag{93}$$

where the matrices are defined as in (68). The corresponding eigenvalue problem

$$[-\omega^2 M + (K + S)]u = 0,$$

or equivalently

$$[-\omega^2 I + M^{-1}(K + S)]u = 0, \tag{94}$$

is associated with distinct left eigenvectors so that

$$v'[-\omega^2 I + M^{-1}(K + S)] = 0 \tag{95}$$

where v represents the left eigenvectors.

A symmetrizable matrix can be expressed as the product of two symmetric matrices one of which is positive definite, and it is not difficult to show that such a matrix is similar to a symmetric matrix [14] via a positive definite transformation matrix. Thus

$$TM^{-1}(K + S)T^{-1} = H_1 \tag{96}$$

where $T = T' > 0$ and $H_1 = H_1'$.

It follows that the eigenvalues of a pseudo-conservative system are all real, and flutter instability is ruled out as in the case of conservative systems. In other words, although the system is nonconservative due to the presence of circulatory forces, which result in the asymmetric matrices E^{α}, the system behaves like a conservative one, stability being lost by divergence only. An example of this class of systems is provided by a simply-supported column which is subjected to combined effect of an axial force and uniformly distributed tangential forces along its length.

Introducing

$$M^{-1}(K + S) = T^{-1}H_1T \tag{97}$$

into (94) and (95) results in the relation

$$v = T^2u \tag{98}$$

between the left and right eigenvectors. (96) can also be written as

$$T^2M^{-1}(K + S) = H_2 \tag{99}$$

where $H_2 = TH_1T$ which is symmetric.

This property of pseudo-conservative systems enables one to make use of the extremum properties of Classical Rayleigh quotient despite the fact that the system is essentially nonconservative. It follows from (94), (98) and (99) that

$$\omega^2 = \frac{\langle u,H_2u\rangle}{\langle u,T^2u\rangle} \tag{100}$$

and

$$\omega_1^2 \le \frac{\langle x, H_2 x\rangle}{\langle x, T^2 x\rangle} \le \omega_n^2 \tag{101}$$

where the frequencies $\omega_1^2, \omega_2^2, \ldots \omega_n^2$ are assumed to be in ascending order.

It is also noted that the eigenvalue problem (94) can be transformed into simple form

$$(-\omega^2 I + H_1)y = 0$$

by pre-multiplying (94) by T and introducing $u = T^{-1}y$, indicating a complete analogy with conservative systems. It then follows that the fundamental characteristic surface and the stability boundary of pseudo-conservative systems possess the convexity properties as in Theorem 2 and its corollary respectively.

Define now the corresponding conservative system as

$$(-\omega^2 M + U + \eta_\alpha E_c^\alpha)u_c = 0 \tag{102}$$

where the matrices E_c^k are the symmetric parts of the originally asymmetric matrices, the skew-symmetric parts being omitted. Based on this concept several lower bound estimates have been established [14, 16, 17]. To give an idea about these general results, consider again a ray in the form $\eta_\alpha = \ell_\alpha \xi$.

The pseudo-conservative, and the corresponding conservative systems are then described by

$$(-\omega^2 M + U - \xi E)u = 0, \text{ where } E = \ell_\alpha E^\alpha, \tag{103}$$

and

$$(-\omega^2 M + U - \xi E_c)u_c = 0 \tag{104}$$

respectively. At a fixed value of ξ, one has

$$\omega_{1c}^2 = \frac{\langle u_c^i, (U - \xi E_c) u_c^i \rangle}{\langle u_c^i, M u_c^i \rangle} \tag{105}$$

where ω_{1c}^2 is the smallest eigenvalue. At the same value of ξ, the smallest eigenvalue of the pseudo-conservative system is given by

$$\omega_1^2 = \frac{\langle u', (U - \xi E) u' \rangle}{\langle u', M u' \rangle} = \frac{\langle u', (U - \xi E_c) u' \rangle}{\langle u', M u' \rangle} \tag{106}$$

which shows that

$$\omega_1^2 \geq \omega_{1c}^2 \tag{107}$$

which shows that the fundamental frequency of the corresponding conservative system provides a lower bound for that of the pseudo-conservative system. For $\xi = 0$, one obtains the natural frequency $\omega_1^2 = \omega_{1c}^2$; as ξ is increased in magnitude, $w_1^2(\xi)$ tends to zero before $\omega_{1c}^2(\xi)$ because of (107), and hence the lower bound property applies to critical loads as well (Figure 6). These results are formulated in the following theorems:

Theorem 4. The fundamental frequency of the corresponding conservative system provides a lower bound for the fundamental frequency of the pseudo-conservative system.

Theorem 5. The critical load of the corresponding conservative system provides a lower bound for the critical load of the pseudo-conservative system.

For several other results concerning the behaviour of pseudo-conservative systems the reader is referred to [14] where the effect of damping is also discussed and will not be presented here.

Identification of the system

Clearly, it is desirable to distinguish a priori a pseudo-conservative system from other nonconservative systems. If the matrix $A = M^{-1}(K + S)$ is symmetrizable all its eigenvalues are real, and for a two-degree-of-freedom system,

$$\text{sign } a_{12} = \text{sign } a_{21}$$

is a sufficient condition for real eigenvalues.

In the general case of n degrees of freedom, one can apply the definition of symmetrizability which is equivalent to

$$DA = H \tag{108}$$

where $D = D' > 0$ and $H = H'$. In other words, the matrix A is symmetrizable if there exist a symmetric positive definite matrix D such that $DA = H$ is symmetric. In many practical situations D can be chosen as a positive diagonal matrix and the problem is then greatly simplified as will be demonstrated in the following example.

Example. Pflüger's column

Consider a simply-supported elastic bar subjected to uniformly distributed tangential forces, η being the intensity of the forces. The governing differential equation of small vibrations is given by [14].

$$sy'''' + \eta(\ell - x)y'' + m\ddot{y} = 0 \tag{109}$$

where $s = EI$, the fexural rigidity, ℓ is the length of the bar and m denotes the mass (constant) per unit length. Primes and dots on the deflection function $y(x,t)$ denote differentiation with respect to x and time respectively. The boundary conditions are

$$\left.\begin{aligned} y(o,t) &= y(\ell,t) = 0 \\ y''(o,t) &= y''(\ell,t) = 0 \end{aligned}\right\} \qquad (110)$$

Introducing

$$y(x) = z(x)e^{i\omega t}$$

into (109) yields

$$sz'''' + \eta(\ell - x)z'' - \omega^2 mz = 0 \qquad (111)$$

with the boundary conditions

$$\left.\begin{aligned} z(o) &= z(\ell) = 0 \\ z''(o) &= z''(\ell) = 0 \end{aligned}\right\} \qquad (112)$$

For the solution of (111), assume a deflection function of the form

$$z = u_1 \sin\frac{\pi x}{e} + u_2 \sin\frac{2\pi x}{e} \qquad (113)$$

$$\equiv u_1\, z_1(x) + u_2\, z_2(x)$$

and apply the Galerkin's procedure to obtain

$$(-M\omega^2 + U - \eta E)u = 0 \qquad (114)$$

where

$$m_{ij} = m\int_o^\ell z_i\, z_j\, dx = \frac{\ell m}{2}\begin{bmatrix} 1 & 0 \\ 0 & 1 \end{bmatrix}$$

$$u_{ij} = s\int_o z''''_i\, z_j\, dx = s\begin{bmatrix} \frac{\pi^4}{2\ell^3} & 0 \\ 0 & \frac{8\pi^4}{\ell^3} \end{bmatrix} \qquad (115)$$

$$e_{ij} = \int_o^\ell (\ell - x)z''_i\, z_j\, dx = \begin{bmatrix} \frac{\pi^2}{4} & \frac{32}{9} \\ \frac{8}{9} & \pi^2 \end{bmatrix}$$

and $u = (u_1, u_2)'$.

To verify that the system described by (114) is pseudo-conservative, it suffices to show that $M^{-1}(U - \eta E)$ is symmetrizable for all values of η. Applying (108), it is first attempted to satisfy (108) with a positive definite diagonal matrix. In this example, it is not difficult to ascertain that the matrix

$$D = \begin{bmatrix} 1 & 0 \\ 0 & 4 \end{bmatrix} \tag{116}$$

does, indeed, satisfy this condition and one obtains

$$H = \frac{1}{m}\begin{bmatrix} s\frac{\pi^4}{\ell^4} - \eta\frac{\pi^2}{2\ell} & -\eta\frac{64}{9\ell} \\ -\eta\frac{64}{9\ell} & s\frac{64\pi^4}{\ell^4} - \eta\frac{8\pi^2}{\ell} \end{bmatrix} \tag{117}$$

and the eigenvalue problem (114) can be written as

$$(-\omega^2 D + H)u = 0 \tag{118}$$

where D and H corresponds to T^2 and H_2 in (99) respectively.

<u>The corresponding conservative</u> system (102) is obtained according to definition where

$$E_c = \begin{bmatrix} \frac{\pi^2}{4} & \frac{20}{9} \\ \frac{20}{9} & \pi^2 \end{bmatrix}.$$

The critical loads of the pseudo-conservative and the corresponding conservative systems can now be determined as $\eta_{cr} = 18.96\ s/\ell^3$ and $\eta^c_{cr} = 18.58\ s/\ell^3$ respectively which indicate that η^c_{cr} is a good lower bound to η_{cr}.

2.5 Gyroscopic systems

A remarkable class of autonomous systems is concerned with gyroscopic forces (e.g. Coriolis force) which result in an equation of the form

$$M\ddot{q} + G\dot{q} + Kq = 0 \tag{119}$$

where $M = M' > 0$, $K = K'$ and $G = -G'$. It is assumed that, besides the conservative forces derivable from a potential, the only velocity dependent forces are of gyroscopic type. Considering the Hamiltonian

$$H = \frac{1}{2} <\dot{q},M\dot{q}> + \frac{1}{2} <q,Kq>$$

one observes that

$$\dot{H} = -<\dot{q},G\dot{q}> = 0.$$

The systems described by (119) are, therefore, called gyroscopic conservative. It also follows that if $K > 0$, the Hamiltonian is a Liapunov function and the system is stable. This is, however, a sufficient condition only, and the positive definiteness of the potential energy (and hence K) is no longer a necessary condition for stability as opposed to conservative systems. In other words, gyroscopic forces may stabilize a conservative system which would have been unstable in their absence. This result follows directly from the stability criterion (74). Another interesting phenomenon associated with gyroscopic conservative systems is that these systems can exhibit flutter instability which was shown to be impossible for "conservative" systems in section 1.3. It can easily be demonstrated that at least two roots of

$$(-\omega^2 M + iG\omega + K)u = 0 \tag{120}$$

coincide when the criticality condition (75),

$$g^2 + 4k = 0, \tag{121}$$

holds for some eigenvector. On the other hand, k = 0 indicates divergence instability and it is clear from (121) that flutter cannot occur before divergence. The significance of flutter instability lies in the fact that it may occur after gyroscopic forces have stabilized an otherwise unstable conservative system, effectively creating flutter regions between the divergence points on the parameter axis. This feature can best be explained by examining the characteristic curves of the system. First note that, due to symmetry of M and K and skew-symmetry of G, the roots of the characteristic equation

$$\Delta(\omega^2) \equiv |-\omega^2 M + i\omega G + K| = 0 \tag{122}$$

appear in pairs $+\omega_k$, $-\omega_k$ $(k = 1,2,\ldots n)$. It follows that the characteristic polynomial resulting from the expansion of (122) is a function of ω^2. The system remains stable as long as all ω_k^2 are real and positive. Furthermore, the divergence points on the parameter axis, corresponding to the vanishing of the ω_k, are the same for both the gyroscopic system and the corresponding non-gyroscopic system which is defined here as the system described by (120) with G = 0. Gyroscopic systems are generally characterized by a parameter ξ, such as the velocity or angular velocity, and G is proportional to ξ while ξ^2 appears in K such that

$$K = U - \xi^2 E, \qquad G = \xi F \tag{123}$$

where $F = -F'$. The characteristic polynomial is a function of ξ^2 which will be referred to as the parameter, and $\xi^2 - \omega^2$ relationship will be explored.

The natural frequencies (for $\xi^2 = 0$) and critical divergence values of ξ^2 are the same for both gyroscopic system and the corresponding G = 0 system. The distinction between the two systems arises in the intervals between such points. The following theorem will now be proved [18]:

Theorem 6. In the region K > 0

$$0 \leq \omega_1^2 \leq (\omega_1^2)_{G=0} \tag{124}$$

where it is assumed that

$$0 < \omega_1^2 \leq \omega_2^2 \leq \ldots. \leq \omega_n^2$$

and

$$0 < (\omega_1^2)_{G=0} \leq (\omega_2^2)_{G=0} \ldots. \leq (\omega_n^2)_{G=0} \tag{125}$$

In other words, the fundamental characteristic curve of the G = 0 system provides an upper bound for that of the gyroscopic system in the region K > 0 (Figure 7).

Proof: When K > 0, the Hermitian matrix $H(\omega) = [-\omega^2 M + i\omega G + K]$ is positive definite for $\omega = 0$ and its smallest eigenvalue λ_1 vanishes for $\omega = \omega_1$, leading to

$$\langle\bar{u}_1, H(\omega_1)u_1\rangle = 0$$

where u_1 is the corresponding eigenvector.

The extremum properties associated with Hermitian matrices then yield

$$\langle v_1, H(\omega_1)v_1\rangle \geq 0 \tag{126}$$

where the real vector v_1 corresponds to $(\omega_1^2)_{G\,=\,0}$ of the non-gyroscopic system

$$(-\omega^2 M + K)v = 0$$

and is assumed to be normalized with respect to M such that

$\langle v_1, Mv_1 \rangle = 1.$

It then follows from (126) that

$$\omega_1^2 \leq (\omega_1^2)_{G=0} \tag{127}$$

since $\langle v_1, Gv_1 \rangle = 0$ and $(\omega_1^2)_{G=0} = \langle v_1, Kv_1 \rangle$.

As ξ^2 is increased beyond ξ_1^2, the fundamental curve may remain in the $\omega^2 > 0$ region, passing the point ξ_1^2 tangentially, or it may enter the instability region $\omega^2 < 0$. In the latter case it has to stay above the fundamental curve of $G = 0$ since the characteristic determinant cannot vanish in that region [14]. The curve normally re-enters the stable region $\omega^2 > 0$ at the second divergent point ξ_2^2 on the load axis. One can then interpret this phenomenon as <u>the stabilizing effect of the gyroscopic forces</u>.

It is possible that the characteristic curve reaches a maximum (Fig. 7) with respect to ξ^2, resulting in flutter instability. This happens when

$$\frac{d(\xi^2)}{d(\omega^2)} = 0 \tag{128}$$

and above this point two roots ω_k^2 become complex conjugate, one of them having positive real part which produces self-excited vibrations with increasing amplitude. (128) is equivalent to (121), and can be used to obtain the critical flutter load directly from the characteristic determinant by observing that

$$\frac{d\xi^2}{d\omega^2} = 0 \text{ leads to } \frac{\partial \Delta(\omega^2, \xi^2)}{\partial \omega^2} = 0 \; . \tag{129}$$

If there are other conservative loading parameters η_α so that the system is described by

$$(-\omega^2 M + i\omega\xi F + U - \xi^2 E - \eta_\alpha E^\alpha)u = 0, \tag{130}$$

the divergence and flutter boundaries are similarly given by

$$\left.\begin{aligned} &\Delta(\omega^2,\xi^2,\eta_\alpha) = 0 \\ &\omega^2 = 0 \end{aligned}\right\} \tag{131}$$

and

$$\left.\begin{aligned} &\Delta(\omega^2,\xi^2,\eta_\alpha) = 0 \\ &\frac{\partial\Delta}{\partial\omega^2} = 0 \end{aligned}\right\} \tag{132}$$

respectively. The <u>initial divergence</u> boundary obviously possesses the convexity properties stated in Corollary 2.1 (see Refs. [19, 14]).

<u>Rotating systems</u>

The eigenvalue problem (130) may describe different systems including fluid-conveying pipes and rotating systems. If attention is restricted to the latter systems, (130) usually takes the special form [19]

$$\left\{-\omega^2\begin{bmatrix} M & 0 \\ 0 & M \end{bmatrix} + 2i\omega\xi\begin{bmatrix} 0 & -M \\ M & 0 \end{bmatrix} + \begin{bmatrix} U & 0 \\ 0 & V \end{bmatrix} - \xi^2\begin{bmatrix} M & 0 \\ 0 & M \end{bmatrix} - \eta\begin{bmatrix} E & 0 \\ 0 & E \end{bmatrix}\right\}\begin{bmatrix} u^1 \\ u^2 \end{bmatrix} = 0 \tag{133}$$

where U and V are associated with two distinct flexural rigidities in two principal directions, U = V indicating a <u>uniform system</u>. For simplicity, it is assumed that there is only one independent conservative load . The characteristic equation now takes the form

$$\begin{vmatrix} U - \eta E - (\xi^2 + \omega^2)M & -2i\omega\xi M \\ 2i\omega\xi M & V - \eta E - (\xi^2 + \omega^2)M \end{vmatrix} = 0. \tag{134}$$

Considering first the uniform system U = V, and introducing a complex displacement [14] into the differential equations leading to (133), one can express (133) simply as

$$(-\omega^2 M - 2\xi\omega M + U - \xi^2 M - \eta E)\, u = 0$$

which upon pre-multiplying by $\bar{u}'$ yields

$$\omega = \frac{-\xi m \pm \sqrt{m(a - \eta e)}}{-m} \tag{135}$$

where $m = \langle \bar{u}, Mu \rangle$, $a = \langle \bar{u}, Uu \rangle$ and $e = \langle \bar{u}, Eu \rangle$.

Clearly, if $\eta = 0$, ω will always be real and flutter instability is then ruled out. Furthermore, divergence is limited to critical points $\xi^2 = a/m$. In the presence of a conservative force η, the flutter boundary $a - \eta e = 0$ consists of straight lines $\eta = \eta_k$ $(k = 1,2,\ldots,n)$ parallel to the ξ^2 axis where η_k are the critical divergence points on η axis. Observing that $U - \eta E$ is positive definite for sufficiently small η, when $a - \eta e = 0$ one has

$$|U - \eta E| = 0 \cdot$$

The flutter frequency is obtained from (135) as $\omega_f = -\xi$. Thus the following Theorems can be stated.

Theorem 7. A uniform rotating system cannot lose stability by flutter if the additional external force η (conservative) is absent.

Theorem 8. The flutter boundary of a uniform rotating system consists of straight lines $\eta = \eta_k$ parallel to the ξ^2 axis where η_k are the divergence points on the η axis.

Next consider the asymmetric system $(U \neq V)$. In many applications V is simply a multiple of U, and the matrices, M, U, and E can be reduced to diagonal forms simultaneously. Then, (134) becomes the product of (2×2) determinants of the form

$$\begin{vmatrix} u - \eta e - (\xi^2 + \omega^2)m & -2i\omega\xi m \\ 2i\omega\xi m & v - \eta e - (\xi^2 + \omega^2)m \end{vmatrix} = 0 \tag{136}$$

where small letters are used to denote the elements of the corresponding matrices. There are n equations like (136) and it can easily be verified that for $\eta = 0$ flutter instability is again ruled out since the discrimant of the characteristic polynomial (in ω^2),

$$(u - v)^2 + 8m\xi^2(u + v),$$

is always positive.

On the other hand, with $\eta \neq 0$, the condition for a double root ω^2 yields the flutter boundary

$$\xi^2 = \frac{(u - v)^2}{8m(2\eta e - u - v)} \tag{137}$$

which describes certain hyperbolas with horizontal asymptote $\xi^2 = 0$ and vertical asymptote

$$\eta = \frac{u + v}{2e} \tag{138}$$

if $e \neq 0$. They are tangential to the divergence boundary at

$$\xi^2 = \frac{u - v}{4m} .$$

Thus the following theorems can be formulated.

Theorem 9. The asymmetric rotating system cannot exhibit flutter instability if the additional external force η is absent.

Theorem 10. The flutter boundary of the asymmetric rotating system consists of branches of hyperbolas with η axis as a horizontal asymptote and with vertical asymptotes midway between the pairs of critical divergence points on the η axis.

Example 1.

Consider a simplified model of a disk (with mass m) mounted on a non-circular weightless shaft which is also subjected to a constant axial compression force η. The shaft is assumed to be rigid with two torsional springs. After simplifications the eigenvalue problem is described by

$$\left\{ -\omega^2\begin{bmatrix}1 & 0\\0 & 1\end{bmatrix} + 2i\omega\xi\begin{bmatrix}0 & -1\\1 & 0\end{bmatrix} + \begin{bmatrix}c_1 - \xi^2 - \eta & 0\\0 & c_2 - \xi^2 - \eta\end{bmatrix}\right\}\begin{bmatrix}u_1\\u_2\end{bmatrix} = 0$$

where c_1 and c_2 represent the elastic rigidities in two principal directions.

Consider first the case $\eta = 0$; the characteristic equation is then given by

$$(\omega^2)^2 - (c_1 + c_2 + 2\xi^2)\omega^2 + (c_1 - \xi^2)(c_2 - \xi^2) = 0, \qquad (139)$$

and in ω^2 - ξ^2 plane one obtains the parabola shown in Fig. 8(i) for $c_1 = 2$ and $c_2 = 6$. Stability is lost by divergence at $\xi^2 = c_1$ and the system remains unstable until $\xi^2 = c_2$. With further increase in angular velocity, stability is regained since the curve then lies in the stable region $\omega^2 > 0$. The characteristic curves for the corresponding conservative system are depicted by dashed lines and in this case they are straight lines. If $c_1 = c_2 = 4$, the system is uniform and the curve touches ξ^2 axis at the only unstable divergence point $\xi^2 = c_1$ (Fig. 8(ii)); otherwise the curve lies totally in the stable region $\omega^2 > 0$. It is further noted that Theorem 6 is demonstrated in both cases.

Flutter instability is not expected for $\eta = 0$ according to theorems established, and it is clear from the figures that the curves cannot, in fact, have a maximum. Now let $\eta \neq 0$; the divergence boundary is readily obtained as

$$\left.\begin{array}{l} c_1 - \xi^2 - \eta = 0 \\ c_2 - \xi^2 - \eta = 0 \end{array}\right\} \qquad (14)$$

which describe two parallel lines (Fig. 9). For the flutter boundary one has the condition

$$\frac{\partial\Delta}{\partial\omega^2} = 2\omega^2 - (c_1 + c_2 + 2\xi^2 - 2\eta) = 0 \qquad (141)$$

which together with (139) yields the flutter boundary

$$1 + 4\xi^2(5 - \eta) = 0 \qquad (142)$$

for $c_1 = 4$ and $c_2 = 6$. This equation describes a hyperbola with asymptotes $\xi^2 = 0$ and $\eta = \frac{4 + 6}{2} = 5$ (Fig. 9). The flutter boundary is tangent to the divergence boundary at $\xi^2 = \frac{6 - 4}{4} = 0.5$ and $\eta = 5.5$. The regions of stability divergence and flutter are indicated by S, D and F respectively in Fig. 9.

Consider now an arbitrary ray τ in the $\xi^2 - \eta$ space defined by $\eta = \ell_1 \tau$ and $\xi^2 = \ell_2 \tau$. Choosing $\ell_1 = \cos 60°$ and $\ell_2 = \cos 30°$, for example, the $\omega^2 - \tau$ relation can be determined readily which is shown to be an ellipse (Fig. 10), demonstrating clearly how flutter can occur. Thus the ray τ, emerging from the origin, intersects the initial divergence boundary first where the system loses stability by divergence. Upon intersecting the second divergence boundary, however, the system regains stability and remains so until the flutter boundary is intersected. At this point the ellipse has a maximum and flutter occurs.

Analysis of elastic flexible shafts lead to analogous results [19].

<u>Example 2: Fluid-conveying pipes</u>

Consider now a circular elastic pipe of mass per unit length m_p conveying a fluid of mass per unit length m_f at constant velocity v. Let x denote the axial coordinate, t time, y(x,t) the transverse displacement $s = EI$ the elastic rigidity and ℓ the length of the pipe. Assuming small displacements, the equation of motion can be established [14, 20] as

$$sy'''' + m_f v^2 y'' + 2m_f v\dot{y}' + (m_f + m_p)\ddot{y} = 0.$$

Define the non-dimensional quantities

$$\xi^2 = \frac{v^2\ell^2 m_f}{\pi^2 EI}, \qquad \omega^2 = \frac{\bar{\omega}^2\ell^4(m_f + m_p)}{\pi^2 EI}, \qquad \gamma = \frac{m_f}{m_f + m_p}$$

where $\bar{\omega}$ is the frequency of vibration.

Using Galerkin's procedure leads to the gyroscopic system (130) with $\eta_\alpha = 0$. If both ends of the pipe are simply supported, one obtains

$$U = \begin{bmatrix} 1 & 0 \\ 0 & 16 \end{bmatrix}, \quad F = \begin{bmatrix} 0 & -\rho \\ \rho & 0 \end{bmatrix}, \quad E = \begin{bmatrix} 1 & 0 \\ 0 & 4 \end{bmatrix} \text{ and } M = \begin{bmatrix} 1 & 0 \\ 0 & 1 \end{bmatrix}$$

with the use of the first two modes of vibration, where $\rho = 256\gamma/9\pi^2$. The characteristic curves for $\rho^2 = 0.5, 1.0, 1.5$ are shown in Figs. 11(i), (ii) and (iii) respectively. The following sequences of behaviours are observed: stability, divergence, flutter, divergence in Fig. 11(i); stability, divergence, stability, flutter in Figs. (ii) and (iii). In each case the second interval of stability on the ξ^2 axis is quite small. The characteristic curves for G = 0 are again straight lines, and here they can be related to a simply-supported column under an axial load of magnitude $m_f v^2$. The upper bound for ω^2 in the first stability region is a good approximation to its actual value in the gyroscopic system.

Standard forms of eigenvalue problems

It has been observed that the characteristic polynomial is a function of ω^2 (and ξ^2). It may be expected, therefore, that it is possible to transform the problem (120) into a standard eigenvalue problem with ω^2 as the eigenvalue. First note that (120) can be represented by

$$(A - \omega I)z = 0 \tag{143}$$

where

$$A = \begin{bmatrix} iM^{-1}G & M^{-1}K \\ I & 0 \end{bmatrix} \text{ and } z = \begin{bmatrix} \omega u \\ u \end{bmatrix}.$$

This problem can in turn be transformed into one involving a Hermitian matrix pencil. Thus, on pre-multiplying (143) by

$$S_1 = \begin{bmatrix} M & 0 \\ 0 & K \end{bmatrix} \tag{144}$$

one obtains

$$(S_2 - \omega S_1)z = 0 \tag{145}$$

where

$$S_2 = \begin{bmatrix} iG & K \\ K & 0 \end{bmatrix},$$

and both S_2 and S_1 are Hermitian. If $K > 0$, so is S_1, and one obtains

$$(S_2 + \omega S_1)S_1^{-1}(S_2 - \omega S\ )z = 0$$

upon pre-multiplying (145) by $(S_2 + \omega S_1)S_1^{-1}$. This equation reduces to

$$(S_3 - \omega^2 S_1)z = 0 \tag{146}$$

where

$$S_3 = S_2\ S_1^{-1}\ S_2 = \begin{bmatrix} K - GM^{-1}G & iGM^{-1}K \\ iKM^{-1}G & KM^{-1}K \end{bmatrix}.$$

Here $S_1 = \bar{S}_1' > 0$, $S_3 = \bar{S}_3'$, and the extremum properties of the Rayleigh Quotient

$$R(x,\bar{x}) = \frac{\langle \bar{x}, S_3\, x\rangle}{\langle \bar{x}, S_1\, x\rangle}$$

is available [21].

In the region where K is no longer positive definite, (144) is not an appropriate choice for S_1. However, one can this time select, for instance,

$$S_1 = \begin{bmatrix} M & 0 \\ 0 & M \end{bmatrix} \tag{147}$$

which results in new S_2 and S_3 as

$$S_2 = \begin{bmatrix} iG & K \\ M & 0 \end{bmatrix}, \quad S_3 = \begin{bmatrix} K - GM^{-1}G & iGM^{-1}K \\ iG & K \end{bmatrix}. \tag{148}$$

In the more explicit form, introducing $K = U - \xi^2 E$ and $G = \xi F$ into (148) and rearranging yields [21] the <u>double eigenvalue problem</u>

$$(-\omega^2 S_1 + S_4 - \xi^2 S_5)p = 0 \tag{149}$$

where

$$S_4 = \begin{bmatrix} U & iFM^{-1}U \\ 0 & U \end{bmatrix}, \quad S_5 = \begin{bmatrix} E + FM^{-1}F & iFM^{-1}E \\ -iF & E \end{bmatrix}.$$

Another interesting form can be generated by pre-multiplying (149) by

$$\begin{bmatrix} I & -iFM^{-1} \\ 0 & I \end{bmatrix},$$

resulting in

$$(-\omega^2 T_1 + T_2 - \xi^2 T_3)p = 0 \tag{150}$$

where

$$T_1 = \begin{bmatrix} M & -iF \\ 0 & M \end{bmatrix}, \quad T_2 = \begin{bmatrix} U & 0 \\ 0 & U \end{bmatrix} \text{ and } T_3 = \begin{bmatrix} E & 0 \\ -iF & E \end{bmatrix}.$$

Effect of damping on the stability boundaries

If internal damping is considered, the differential equations of the motion take the form

$$M\ddot{q} + (D + G)\dot{q} + Kq = 0 \tag{151}$$

where $D \geq 0$. According to the KTC theorem (Kelvin-Tait-Chetaev) an equilibrium position which is stable (unstable) under purely potential forces ($D = G = 0$ in (151)), remains stable (unstable) with the addition of gyroscopic and dissipative forces. If $D > 0$ or $D \geq 0$ but pervasive damping prevails (as in the case of conservative systems), the dissipative system is asymptotically stable if $K > 0$. In other words, all the higher regions of stability and the stabilizing effect of gyroscopic forces are wiped out by dissipation except for the fundamental region of stability. For the proof of this theorem see for example [14].

Introduction of external damping or circulatory forces into the system results in [14] an equation of the form (68), and the general criterion of stability (73) applies as long as $D \neq 0$. If $D = 0$ the criterion (74) must be used. External damping introduces an additional non-negative definite matrix D_e so that $D = D_i + D_e$ where internal damping is now denoted by D_i. First, note that if $D = 0$, one obtains an important result on the basis of (74): A non-dissipative gyroscopic system is in general unstable if circulatory forces are also present. Next, writing the stability criterion (73) in the form

$$d > 0, \qquad kd^2 + \left(\frac{dg}{2}\right)^2 - \left(\frac{dg}{2} - s\right)^2 > 0 \tag{152}$$

one deduces that a gyroscopic system which is unstable under internal damping may be stabilized with the addition of external damping.

Consider now a rotating system. The governing equations of such a system take the special form [22]

$$\begin{aligned} M\ddot{p} + (d_i + d_e)M\dot{p} - 2\xi M\dot{q} + (U - \eta E - \xi^2 M)p - d_e\xi Mq &= 0 \\ M\ddot{q} + (d_i + d_e)M\dot{q} + 2\xi M\dot{p} + (V - \eta E - \xi^2 M)q + d_e\xi Mp &= 0 \end{aligned} \tag{153}$$

where d_i and d_e are the coefficients of internal and external damping respectively.

If the system is uniform $(U = V)$, these equations can be combined into

$$M\ddot{r} + (d_i + d_e)M\dot{r} + 2i\xi M\dot{r} + (U - \eta E - \xi^2 M)r + id_e\xi Mr = 0 \tag{154}$$

upon introducing the complex displacement $r = p + iq$. Assuming $r = u\,e^{i\omega t}$ and pre-multiplying by $\bar{u}'$ yields

$$-\omega^2 m + i(d_i + d_e)\omega m - 2\omega\xi m + (a - \eta e - \xi^2 m) + id_e\xi m = 0$$

which at the onset of flutter (ω real) results in

$$\begin{aligned} &-\omega^2 m - 2\omega\xi m + a - \eta e - \xi^2 m = 0 \\ \text{and}\quad &\omega(d_i + d_e)m + d_e\xi m = 0 \end{aligned} \tag{155}$$

where $m = \langle\bar{u},Mu\rangle$, $a = \langle\bar{u},Uu\rangle$, etc. as before.

Eliminating ω between these equations yields

$$a - \eta e - \left(\frac{d_i}{d_i + d_e}\right)\xi^2 m = 0$$

which implies

$$\left|U - \eta E - \left(\frac{d_i}{d_i + d_e}\right)\xi^2 M\right| = 0, \tag{156}$$

defining the flutter boundary. This equation is essentially similar to (77) and the convexity properties contained in Theorem 2 applies here as well. Hence, one has the following result.

Theorem 11. The initial flutter boundary of a uniform rotating system cannot have convexity toward the origin of the (ξ^2,η) plane.

Divergence boundary consists of the critical points on the η axis provided $d_e \neq 0$, as can readily be seen from (155) by setting $\omega = 0$. If $d_e = 0$, the KTC theorem applies and the divergence boundary is indentical to that of undamped system. On the other hand, if $d_i = 0$, the flutter boundary takes the form of straight lines $\eta = \eta_j$ given by $(U - \eta E) = 0$ which are recognized as the flutter boundary of the undamped system. In view of (156) one can then state the following result:

Theorem 12. The flutter (divergence) boundary of the undamped system provides an upper (lower) bound for the flutter boundary of the damped system.

An increase in d_e enlarges the region of stability while an increase in d_i reduces it. In Fig. 12, the flutter boundary of a simply-supported rotating shaft which is subjected to axial force η is shown. The boundary has been determined through a two-mode Galerkin procedure [19].

In the asymmetric case $(U \neq V)$, the divergence boundary is obtained from the characteristic equation associated with (153) as

$$\begin{vmatrix} U - \eta E - \xi^2 M & -d_e \xi M \\ d_e \xi M & V - \eta E - \xi^2 M \end{vmatrix} = 0 \qquad (157)$$

where diagonal matrices are positive definite in the fundamental region of stability of the $d_e = 0$ system. Since (157) cannot hold in this region [14] one formulates the following theorem:

Theorem 13. The initial divergence boundary of the undamped system is a lower bound to that of the damped system.

It can be shown [14] that if the matrices M, $U(= \alpha V)$ and E are simultaneously diagonalizable the following holds

Theorem 14. The flutter boundary of the undamped system is an upper bound for that of the damped system.

Fig. 13 illustrates the boundaries of an asymmetric shaft which was analyzed by Galerkin's procedure [19].

2.6 Circulatory systems

As in the case of pseudo-conservative systems, the differential equations of motion are given by

$$M\ddot{q} + (U - \eta_\alpha E^\alpha)q = 0 \tag{158}$$

where some E^α is asymmetric. Here, however, the condition that the matrix $M^{-1}(U - \eta_\alpha E^\alpha)$ remains symmetrizable for all values of η_α is removed. The eigenvalues can be complex and, on the basis of the eigenvalue problem

$$\begin{aligned} &[-\omega^2 M + (K + S)]u = 0 \\ &v'[-\omega^2 M + (K + S)] = 0, \end{aligned} \tag{159}$$

they can generally be expressed as the stationary values of the generalized Rayleigh quotient

$$R(x,y) = \frac{\langle y,(K + S)x\rangle}{\langle y,Mx\rangle}, \tag{160}$$

where x and y are arbitrary vectors.

It follows from (74) that the circulatory system (159) is stable if and only if $k > 0$ and $s = 0$ for all eigenvectors u normalized with respect to M.

Alernatively, observing that the matrix $M^{-1}(K + S)$ remains symmetrizable as long as ω^2 are real and distinct, the following results can immediately be stated:

Theorem 15. A circulatory system is stable if and only if there exist a symmetric positive definite matrix H such that $HM^{-1}(K + S)$ is symmetric and positive definite

Corollary. If the matrix $HM^{-1}(K + S)$ is symmetric, flutter instability is ruled out.

Pre-multiplying the first equation in (159) by M^{-1} and H, it is noted that

$$[-\omega^2 H + HM^{-1}(K + S)]u = 0$$

yields real positive roots ω^2 under the conditions of Theorem 15, and real roots if its corollary applies.

At a flutter point where two roots ω_1^2 and ω_2^2 coalesce, the matrix $M^{-1}(K + S)$ is no longer symmetrizable and the eigenvectors are not linearly independent. At this point the matrix $M^{-1}(K + S)$ is defective [14].

The flutter and divergence boundaries are given by

$$\left.\begin{aligned} &\Delta(\omega^2,\eta_\alpha) \equiv |-\omega^2 M + U - \eta_\alpha E^\alpha| = 0 \\ &\frac{\partial\Delta}{\partial\omega^2} = 0 \end{aligned}\right\} \tag{161}$$

and

$$|U - \eta_\alpha E^\alpha| = 0 \tag{162}$$

respectively. An alternative flutter condition can be obtained by differentiating the first of (159) with respect to ω^2, pre-multiplying by v' and using $\partial\eta_\alpha/\partial\omega^2 = 0$ as

$$\langle v,Mu\rangle = 0 \,. \tag{163}$$

This condition leads to certain max-min properties of the generalized Rayleigh quotient (160) in the case of two-degree-of-freedom systems which are in contrast with those of classical Rayleigh quotient [23]. Thus, while the latter quotient yields values between the smallest and greatest eigenvalues, the former gives values outside this region [23]. In other words, while Rayleigh-Ritz procedure for conservative systems leads to an upper bound to the smallest eigenvalue, an analysis based on left and right eigenvectors as defined here will result in a lower bound [23].

This property can be used to demonstrate that the stability boundary of circulatory systems can have convexity properties in contrast with the conservative systems. Considering

$$\langle v,(-\omega^2 M + U - \eta_\alpha E^\alpha)u\rangle = 0 \tag{164}$$

obtained from (159), at a flutter point A where the left and right eigenvectors are x and y respectively, one has

$$\langle y,(U - \eta_\alpha E^\alpha)x\rangle = 0 \tag{165}$$

since the flutter condition $\langle y,Mx\rangle = 0$ holds at this point. It can be shown [23] that (165) describes a plane tangent to the flutter boundary. Now consider a ray defined by $\eta_\alpha = \ell_\alpha \eta$ which intersects the flutter boundary initially at $\eta = \eta_1 > 0$ and the tangent plane (165) at $\eta = \eta_t$ (Fig. 14). In view of the flutter condition (163) one then has

$$\eta_1 = \frac{\langle v',Uu'\rangle}{\langle v',Eu'\rangle} = R(u',v')$$

and

$$\eta_t = \frac{\langle y,Ux\rangle}{\langle y,Ex\rangle} = R(x,y)$$

where $E = \ell_\alpha E^\alpha$ and u', v' are the corresponding eigenvectors. It is observed that the generalized Rayleigh quotient gives the distance η_t from the origin to the intersection of the ray with the tangent plane while its stationary value R(u',v') is the distance η_1 to the flutter boundary. According to extremum properties of R(x,y) one then has

$$\eta_t \leq \eta_1$$

for a two-degree-of-freedom system. The following result can then be stated:

Theorem 16. The initial flutter boundary of a circulatory system with two degrees of freedom cannot have concavity toward the origin.

An alternative method to prove Theorem 16 as well as a similar theorem for divergence boundary is given in [24, 14], and the result, thus covering the entire stability boundary, can be of practical use in obtaining lower and upper bound estimates.

Some typical examples are shown in Fig. 15 whose boundaries are depicted in Figs. 16 - 18. The divergence (flutter) boundary denoted by DB (FB) and the stability boundary is shaded. Example (a) is of a pseudo-conservative system, and it is included here for comparison. Example (b) and (c) were first examined in [25] and [26] respectively. The results shown here are obtained through a two-mode Galerkin's procedure [24].

Effect of velocity dependent forces

Suppose the velocity dependent forces are associated with an asymmetric matrix and that both D and G are present in the governing equations which is then in the form (68) and the general stability criterion (73) applies. In case D = 0, it was shown in section 2.5 that the system is generally unstable. If the velocity-dependent forces are small so that one has

$$[\lambda^2 M + \lambda\varepsilon(D + G) + (K + S)]u = 0$$

where ε is small, it can be shown that the presence of such forces may have a destabilizing effect [27]. It is first observed that in the region of stability $\varepsilon = 0$ system has real eigenvectors. If one considers then the eigenvectors u in the form $u = u_r + i\varepsilon u_I$ where u_r belongs to $\varepsilon = 0$ system, the stability criterion (73) reduces to

$$d_1 > 0 \qquad k_1 d_1^2 - s^2 > 0$$

as a first order approximation ($\varepsilon \to 0$). Here d_1 and k_1 are in terms of the real vectors u_r just as in the $\varepsilon = 0$ case. Since the undamped system ($\varepsilon = 0$) is stable if and only if $k_1 > 0$ and $s = 0$, <u>it follows that introduction of velocity dependent forces may have a destabilizing effect</u>. It is immediately noted, however, that this is not always the case, and velocity-dependent forces may not have a destabilizing effect. It can readily be demonstrated, for example, that if the velocity-dependent forces are associated with the matrix εM, and the $\varepsilon = 0$ system is stable, the $\varepsilon \neq 0$ system is also stable [14].

Figures, Part II

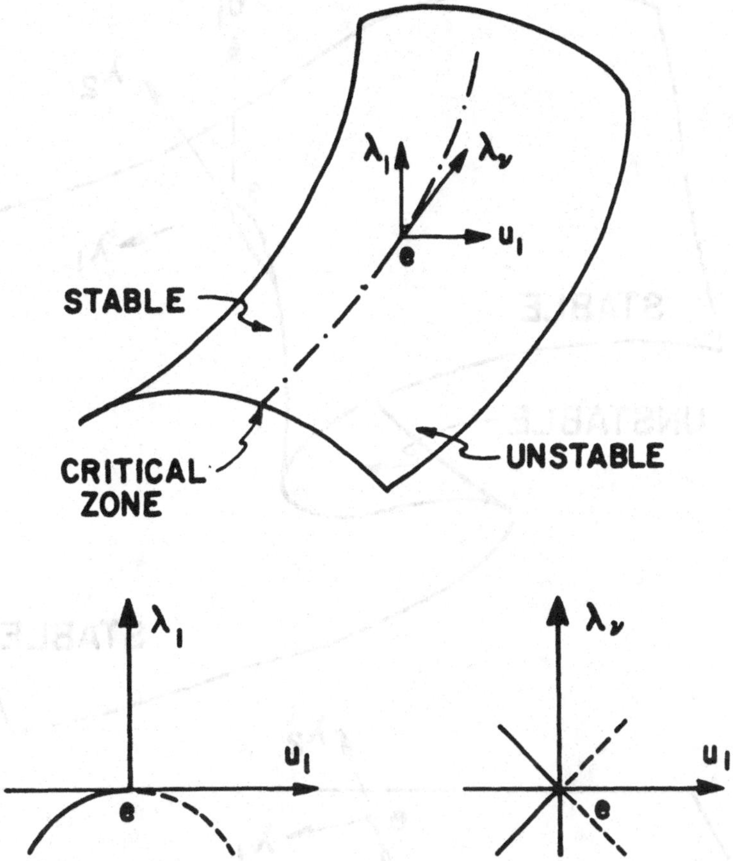

Figure 1

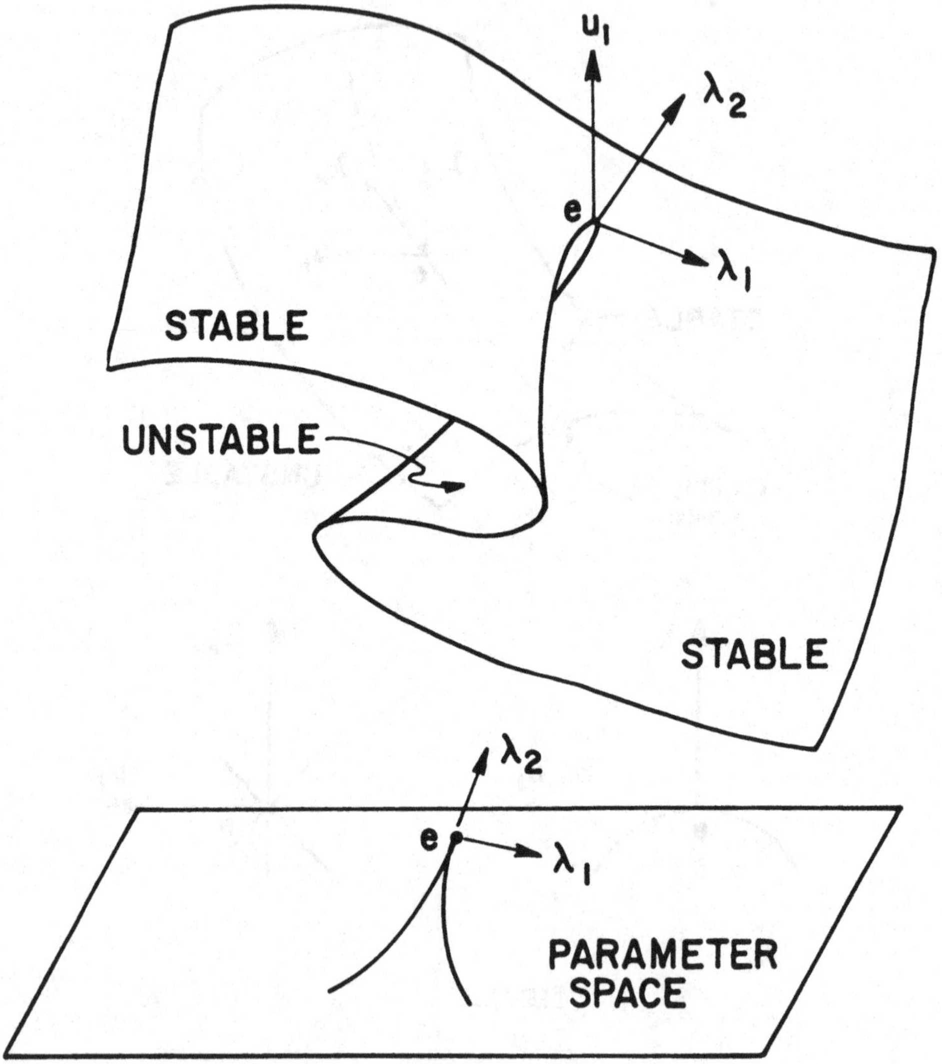

Figure 2

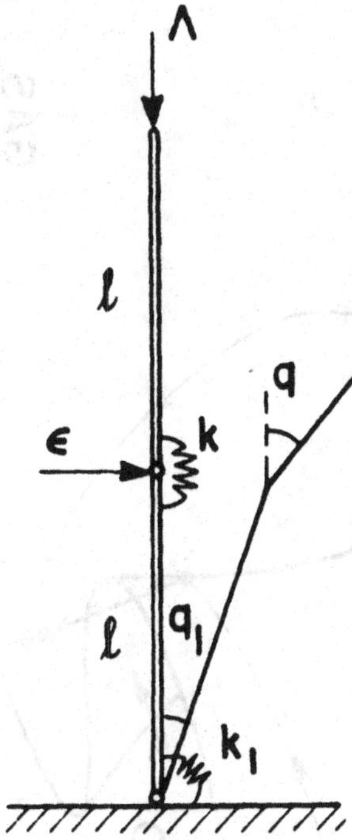

Figure 3

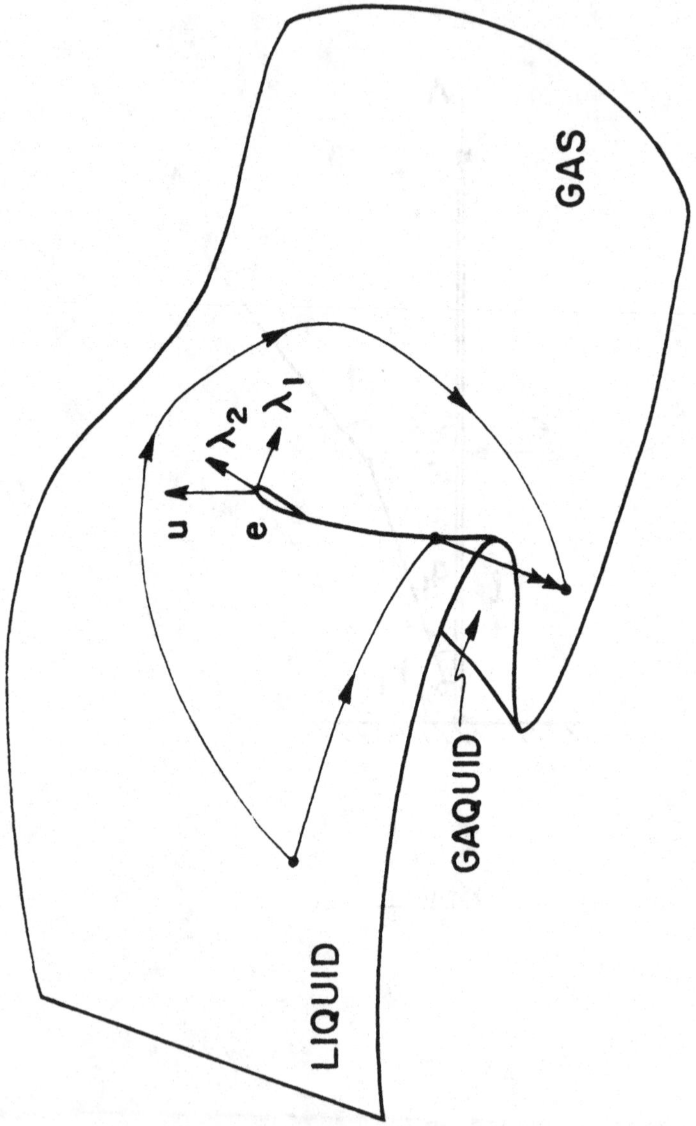

Figure 4

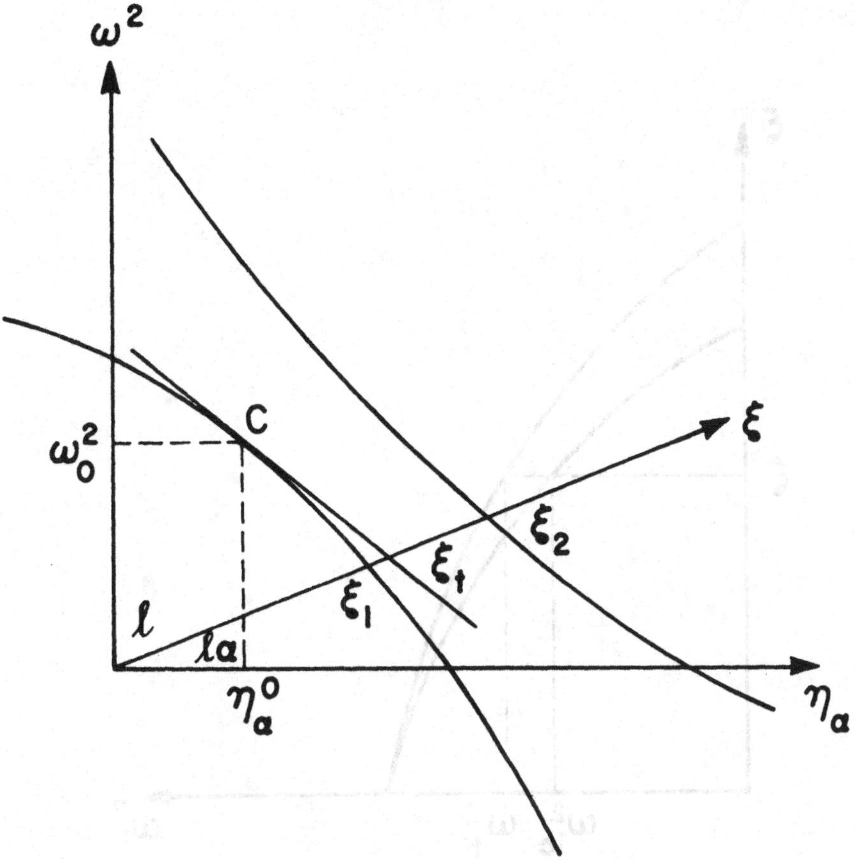

Figure 5

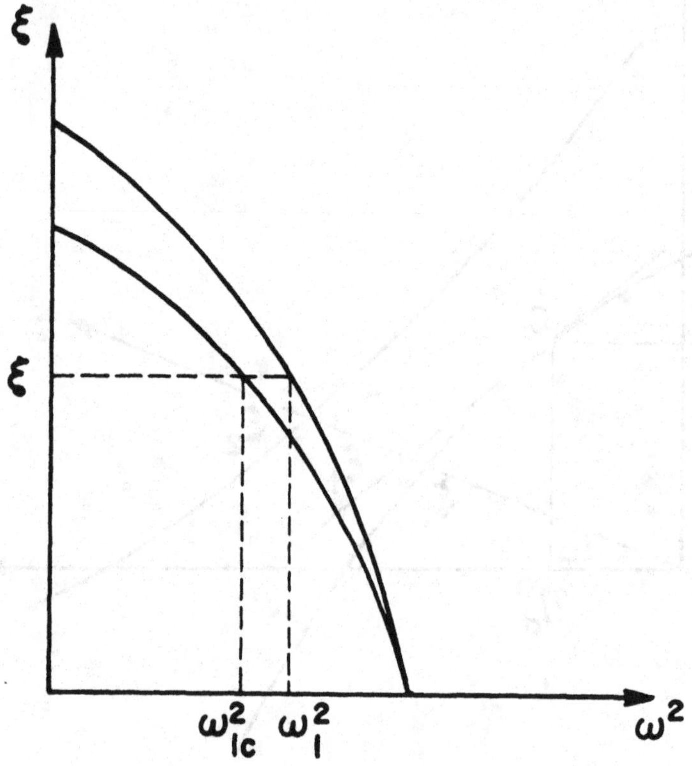

Figure 6

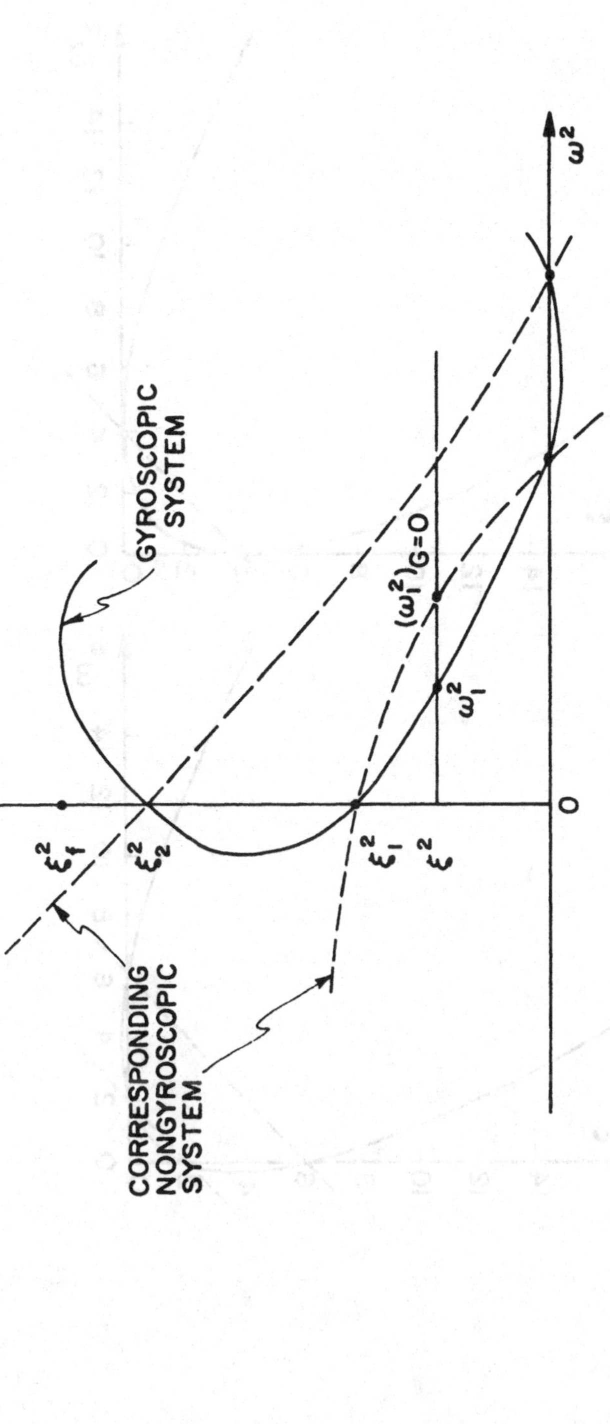
GYROSCOPIC SYSTEM
CORRESPONDING NONGYROSCOPIC SYSTEM
ε^2
ω^2
ε_f^2
ε_2^2
ε_1^2
ε^2
0
$(\omega_1^2)_{G=0}$
ω_1^2

Figure 7

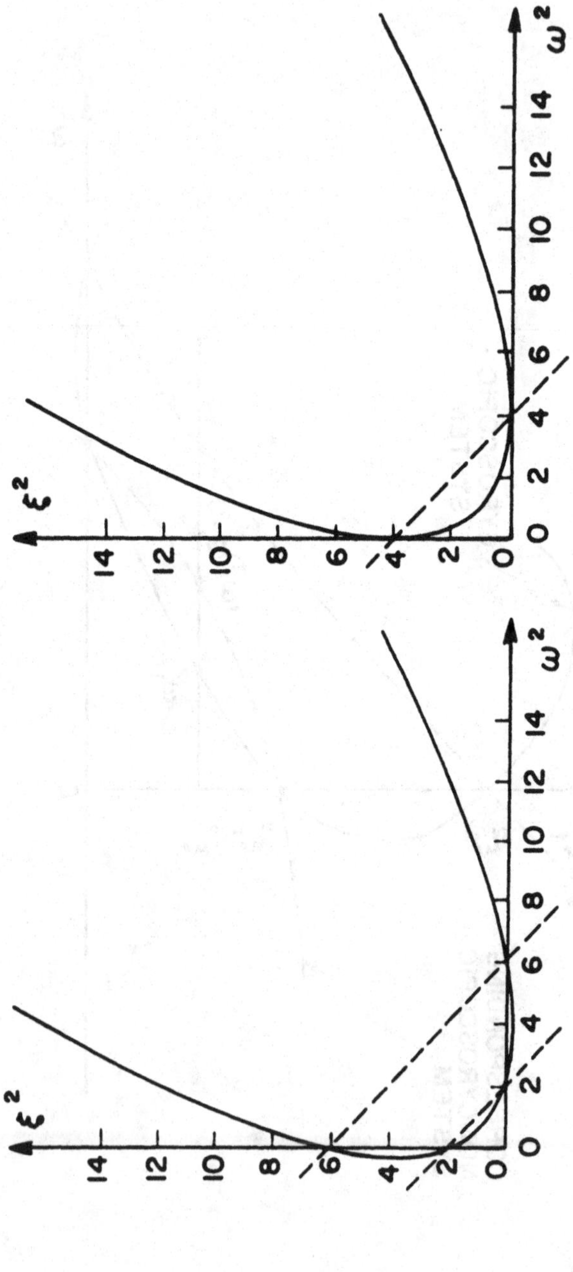

ε^2
ω^2
0
2
4
6
8
10
12
14
ε^2
ω^2
0
2
4
6
8
10
12
14

Figure 8

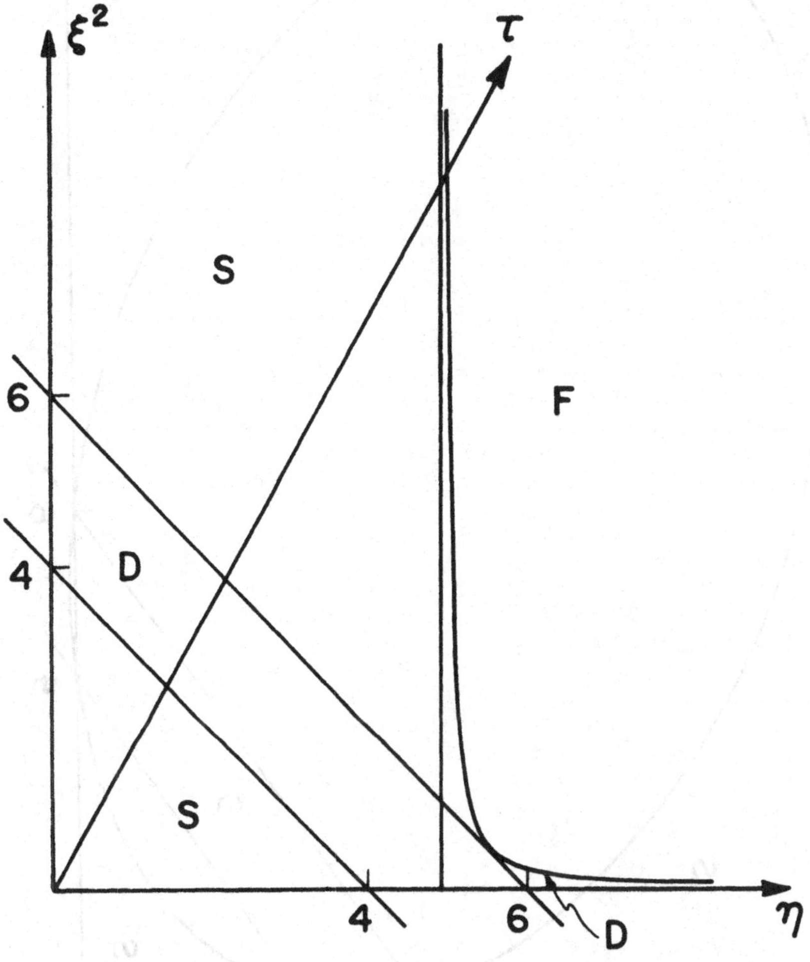

Figure 9

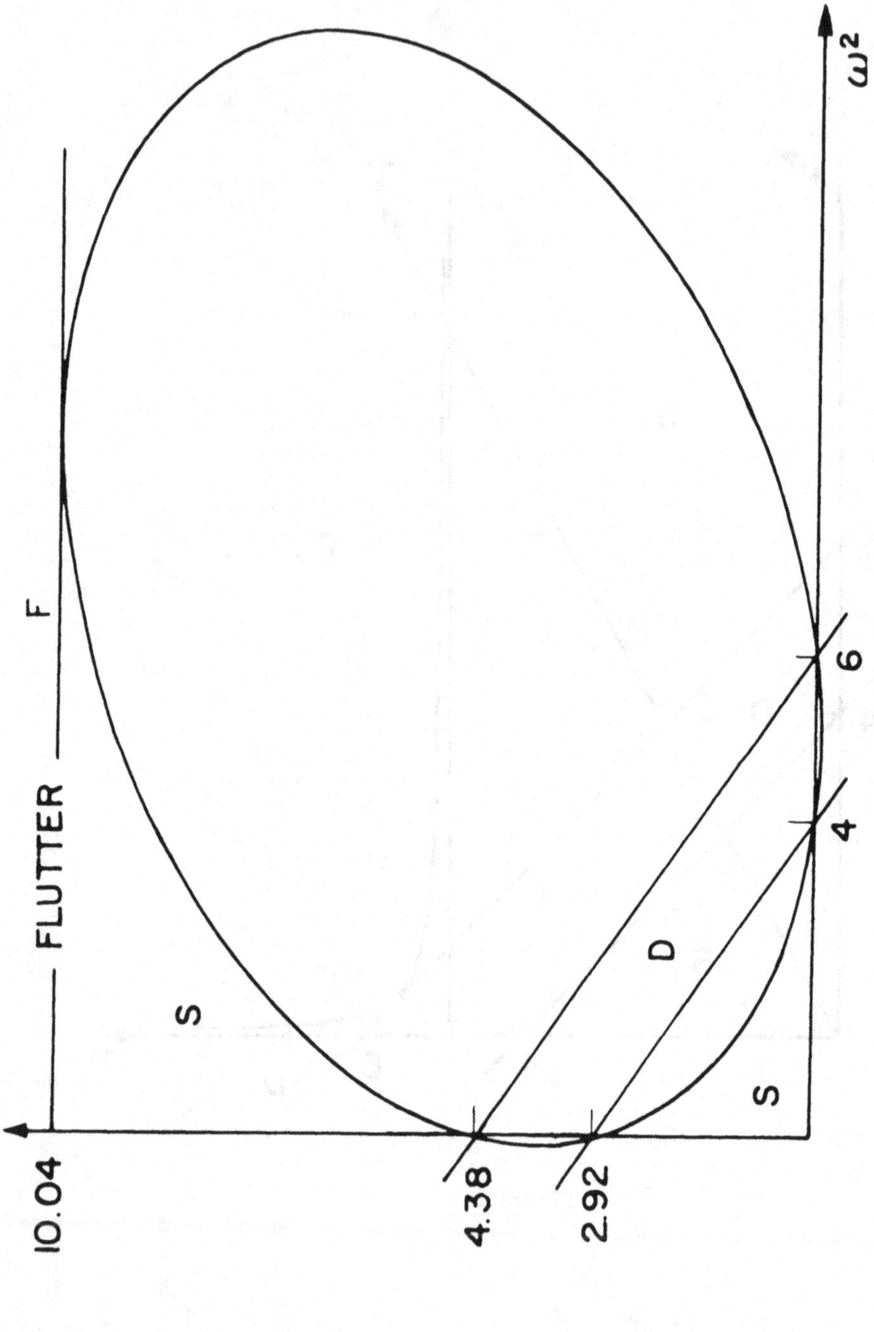

Figure 10

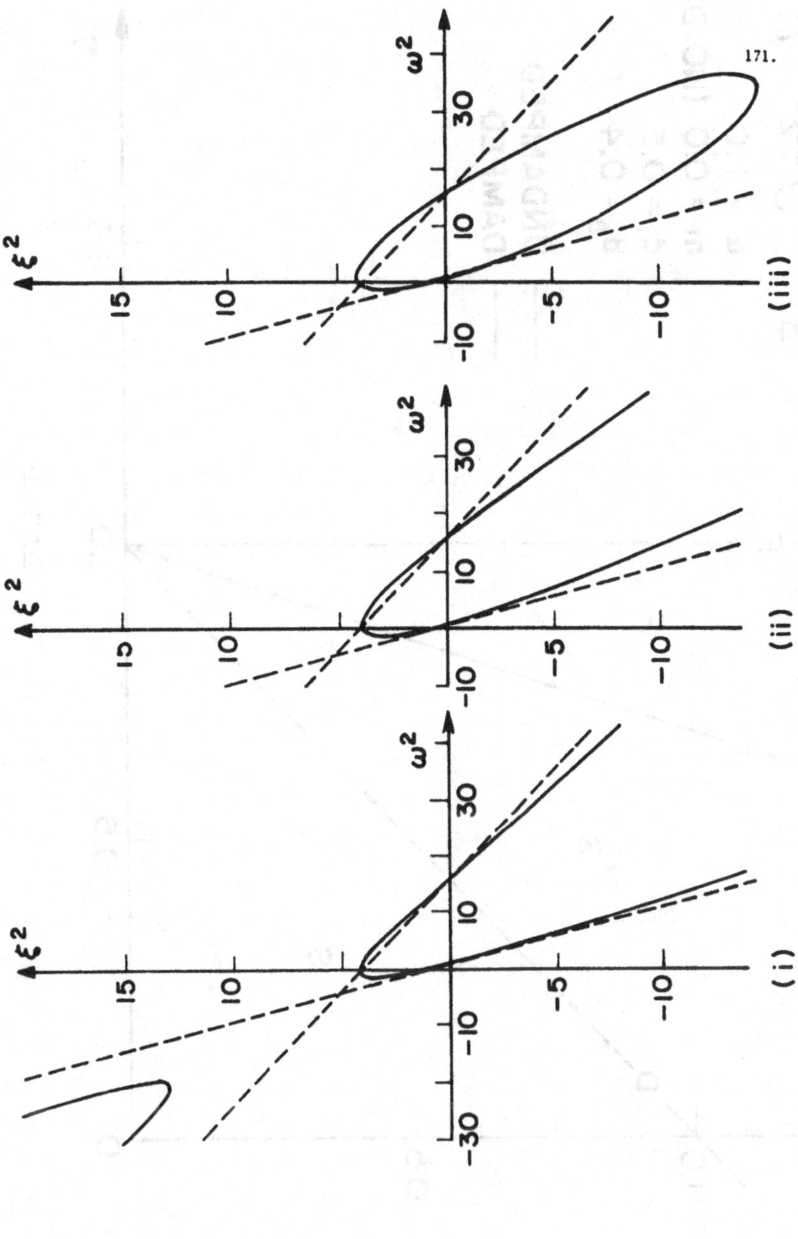

Figure 11

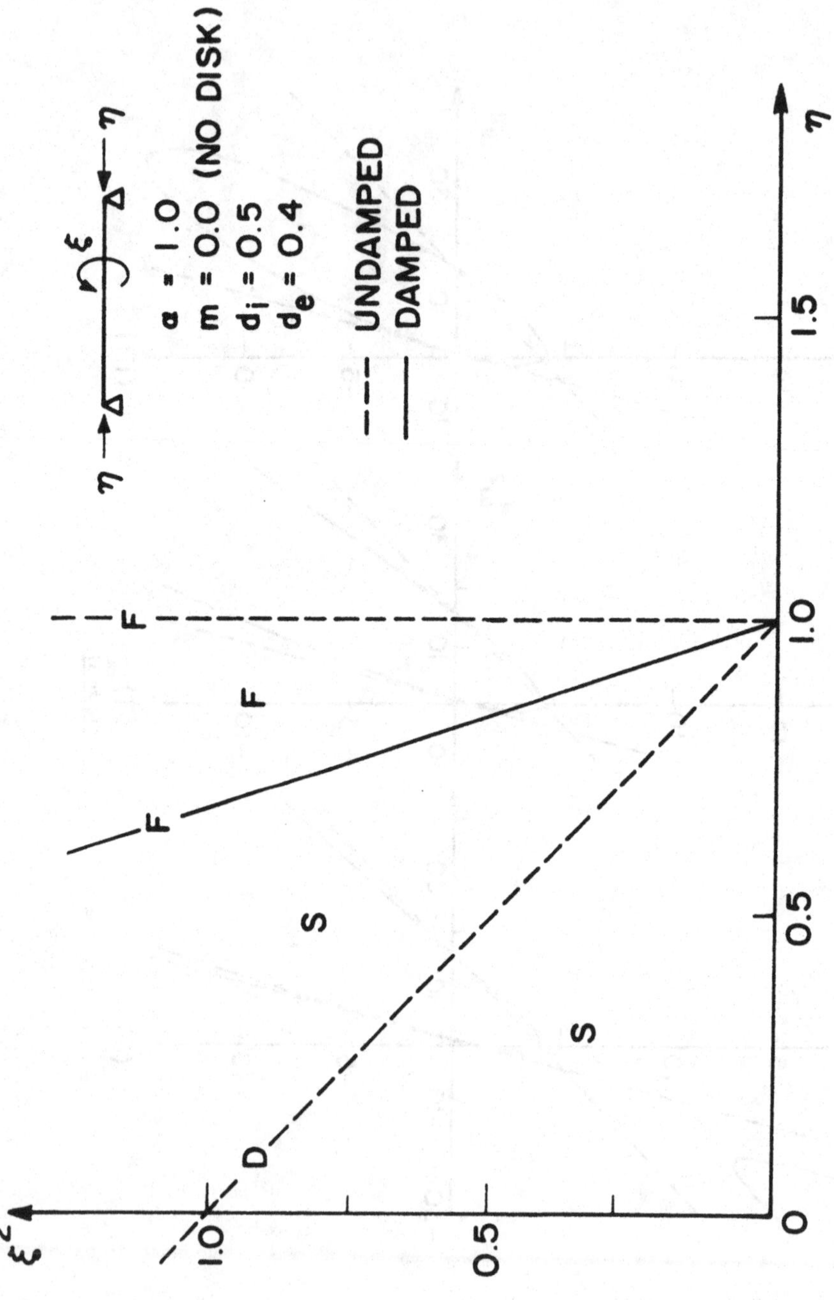
α = 1.0
m = 0.0 (NO DISK)
d_i = 0.5
d_e = 0.4
UNDAMPED
DAMPED
η
ξ
F
F
F
S
S
D
ξ^2
1.0
0.5
0
0.5
1.0
1.5
η

Figure 12

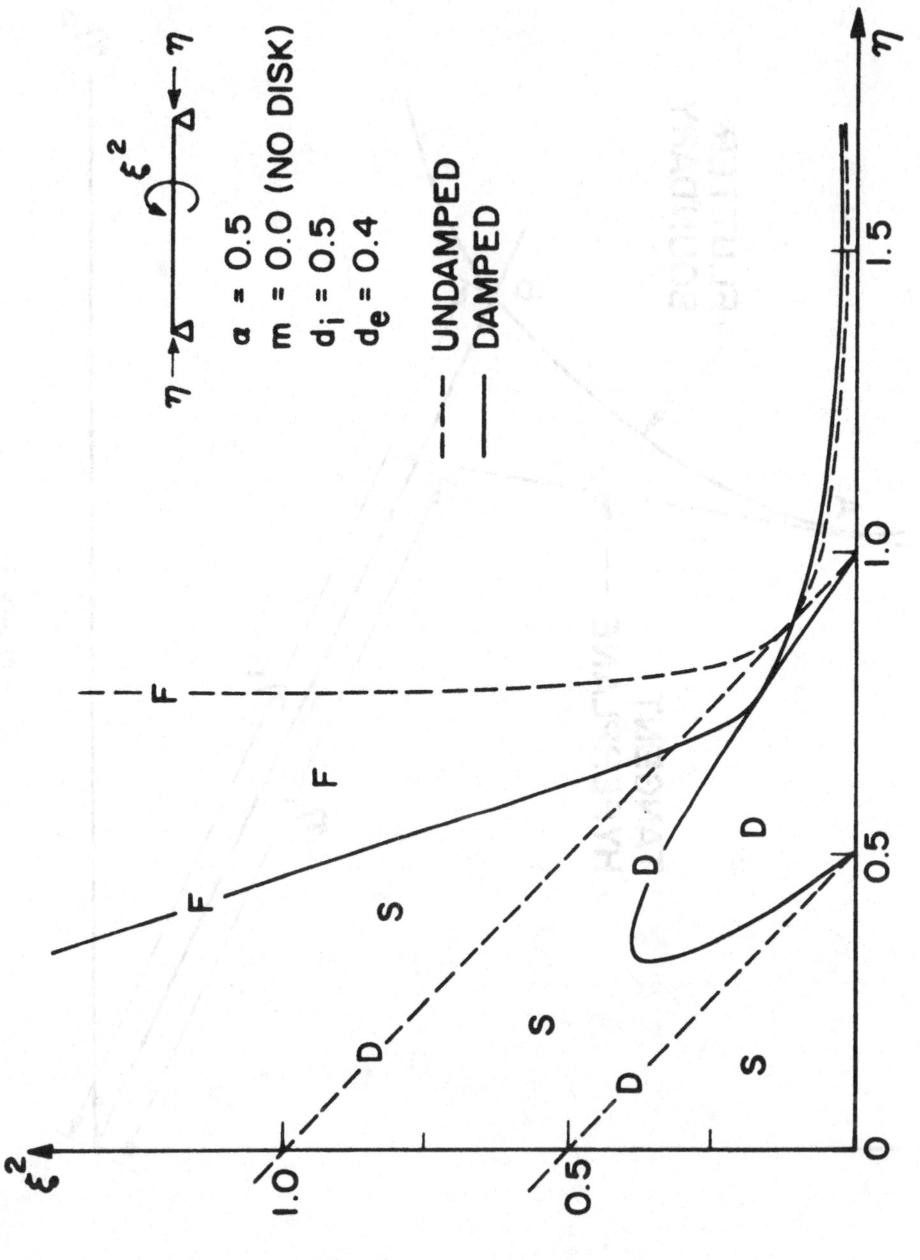
η
ε²
α = 0.5
m = 0.0 (NO DISK)
d_i = 0.5
d_e = 0.4
UNDAMPED
DAMPED
F
F
F
S
S
S
S
D
D
D
D
D
0
0.5
1.0
1.5
0.5
1.0
ε²
η

Figure 13

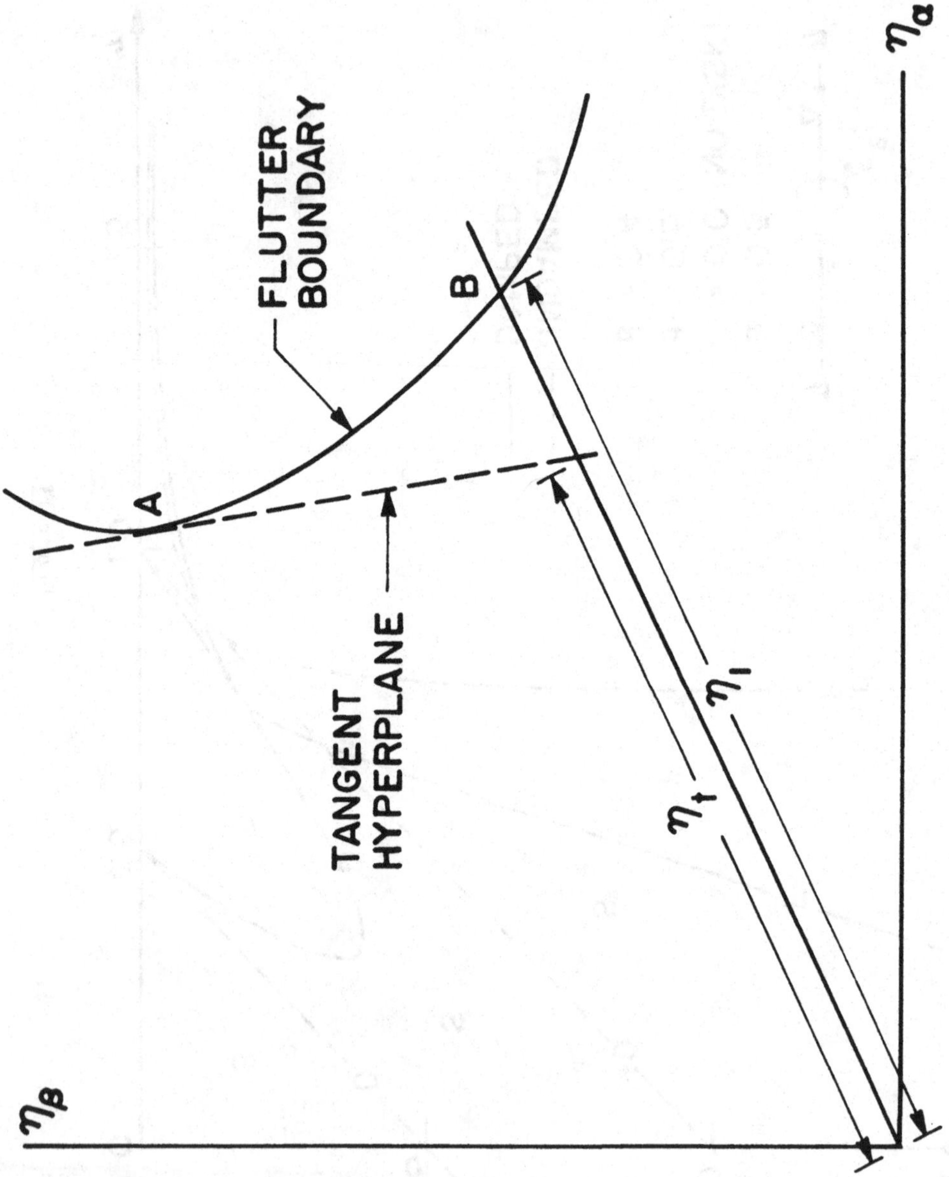
FLUTTER BOUNDARY
TANGENT HYPERPLANE
A
B
η_β
η_α
η_t
η_1

Figure 14

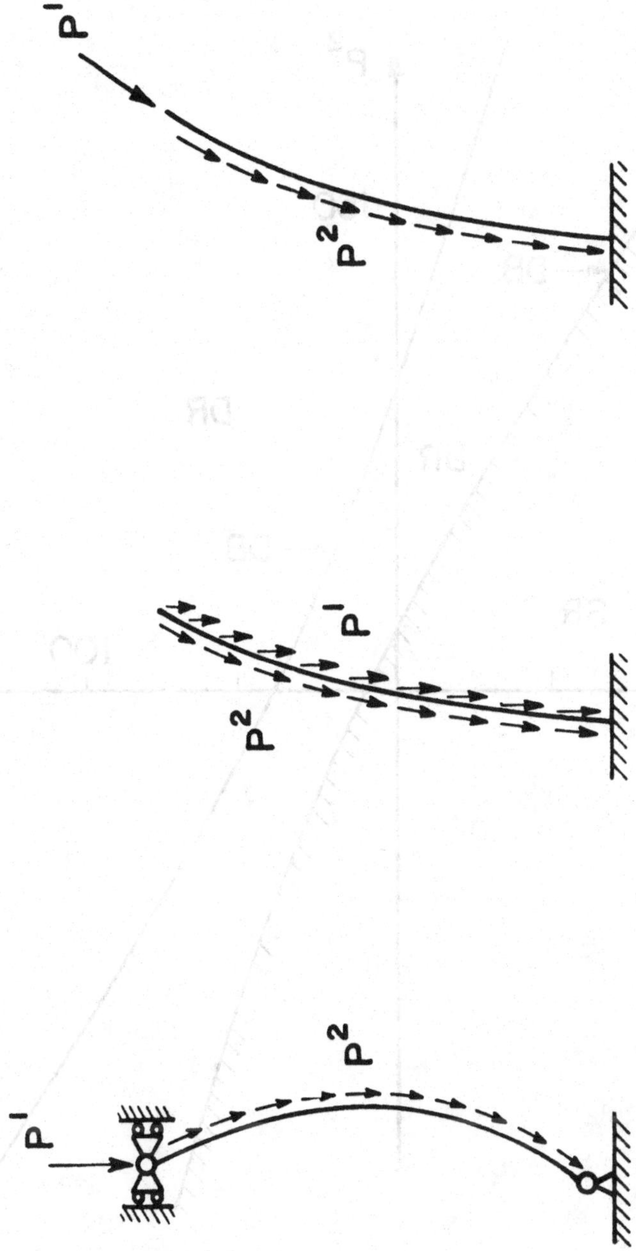

Figure 15

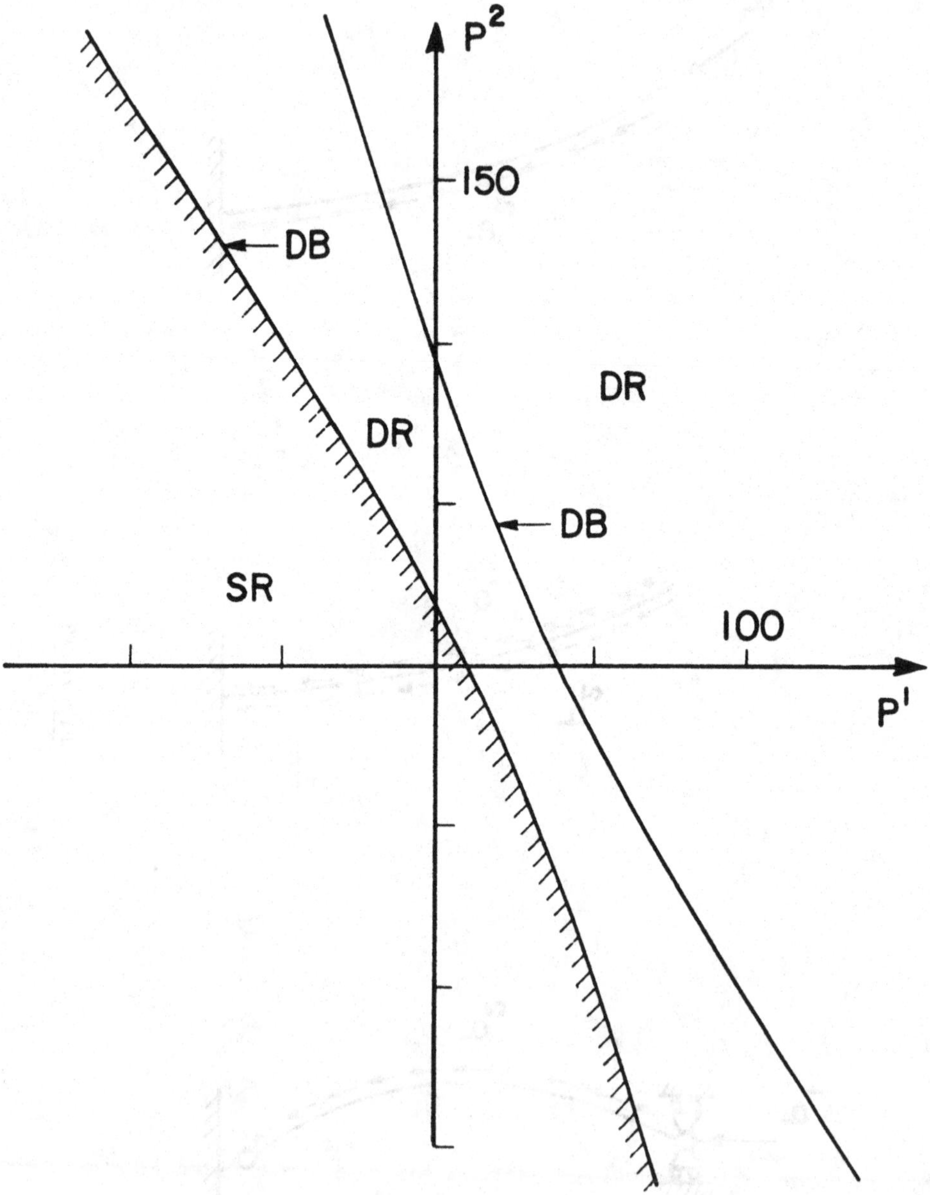

Figure 16

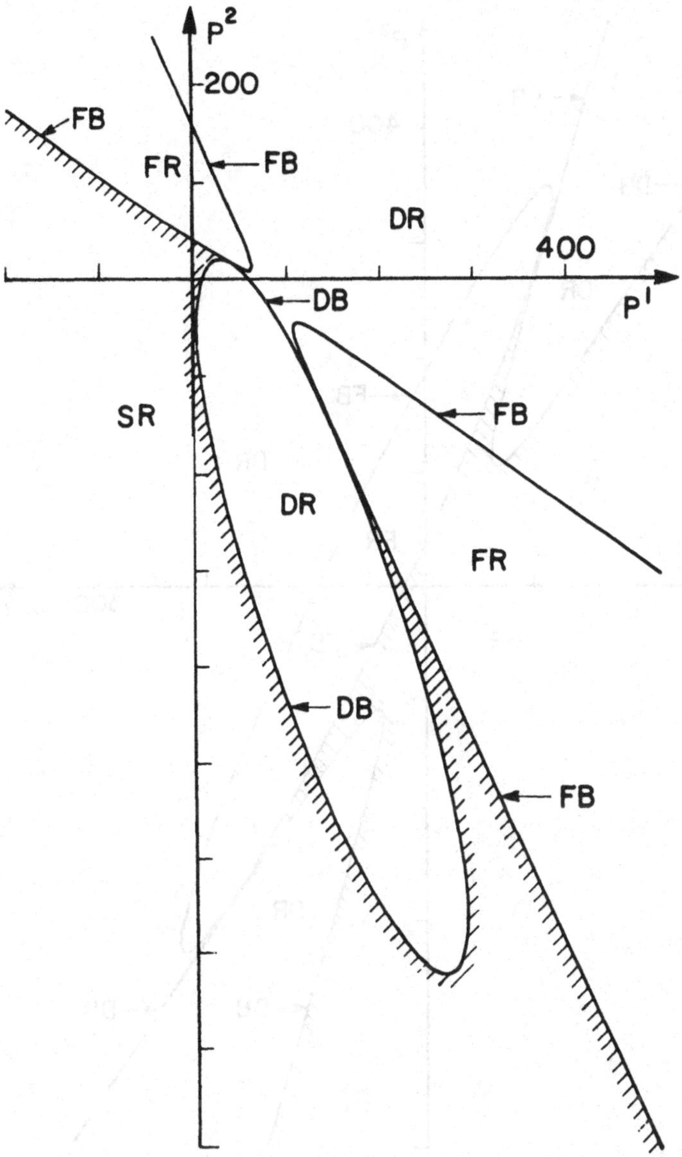

Figure 17

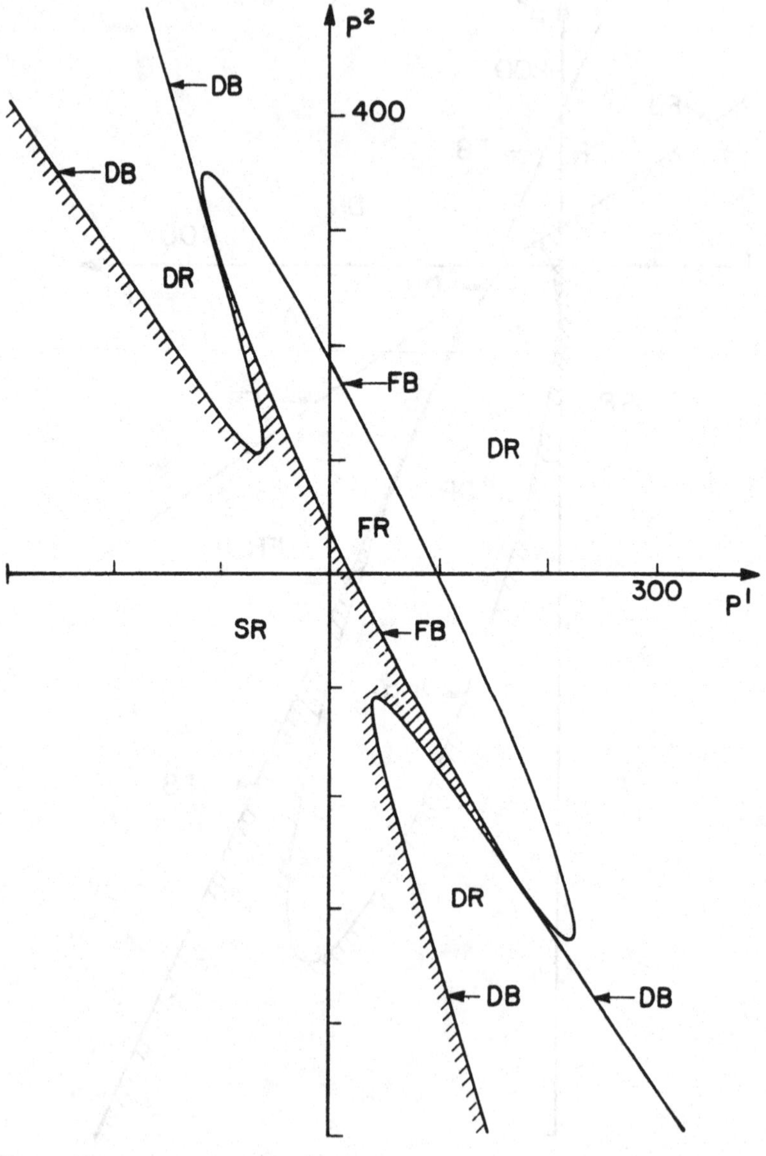

Figure 18

References, Part II

[1] KOITER, W. J., "On the Stability of Elastic Equilibrium", Dissertation, Holland, 1945.

[2] THOMPSON, J. M. and HUNT, G. W., "A General Theory of Elastic Stability", John Wiley & Sons, 1973.

[3] ROORDA, J., "Concepts in Elastic Structural Stability", Mechanics Today (ed. by N. Nasser), 1, Pergamon, 1972.

[4] HUSEYIN, K., "Elastic Stability of Structures Under Combined Loading", Ph.D. Thesis, University of London, 1967.

[5] HUSEYIN, K., "Nonlinear Theory of Elastic Stability", Noordhoff International Publishing, 1975.

[6] THOM, R., "Structural Stability and Morphogenesis", Translated by D. H. Fowler, Benjamin, Inc., 1975.

[7] HUSEYIN, K., "The Multiple-Parameter Stability Theory and Its Relation to Catastrophe Theory", Proc. Problem Analysis in Science and Engineering (ed. by F. Branin and K. Huseyin), Academic Press, 1977.

[8] HUSEYIN, K. and MANDADI, V., "Classification of Critical Conditions in the General Theory of Stability", Mech. Res. Comm. 4, 1977, pp. 11-15.

[9] MANDADI, V., "On the General Theory of Stability and Its Applications", Ph.D. thesis, University of Waterloo, 1976.

[10] HUSEYIN, K., "The Multiple-Parameter Perturbation Technique for the Analysis of Nonlinear Systems", Int. J. of Nonlinear Mech. 8, 1973, pp. 431-443.

[11] MANDADI, V. and HUSEYIN, K., "On the Nonlinear Stability Analysis of Gradient Systems", to be published.

[12] FOWLER, D. H., "The Riemann-Hugoniot Catastrophe and Van der Waals' Equation", Towards a Theoretical Biology (ed. by C. W. Waddington), 4, 1972, pp. 1-7.

[13] LASALLE, J. and LEFSCHETZ, S., "Stability by Liapunov's Direct Method", Academic Press, New York, 1961.

[14] HUSEYIN, K., "Vibrations and Stability of Multiple Parameter Systems", Noordhoff International Publishing (1977).

[15] LEIPHOLZ, H. H. E., "Stability Theory", Academic Press, Inc. New York and London, 1970.

[16] HUSEYIN, K. and LEIPHOLZ. H. H. E., "Divergence Instability of Multiple-Parameter Circulatory Systems", Quart. Appl. Math. 31(2), 1973, pp. 185-197.

[17] HUSEYIN, K. and PLAUT, R. H., "Application of the Rayleigh Quotient to the Eigenvalue Problems of Pseudo-Conservative Systems", J. Sound Vib. 33(2), 1974, pp. 201-210.

[18] HUSEYIN, K. and PLAUT, R. H., "Transverse Vibrations and Stability of Systems with Gyroscopic Forces", J. Struct. Mech. 3(2), 1974-75, pp. 163-177.

[19] HUSEYIN, K. and PLAUT, R. H., "Divergence and Flutter Boundary of Systems Under Combined Conservative and Gyroscopic Forces", Proc. IUTAM Symp. Dynamic of Rotors (ed. Niordson), Springer-Verlag, 1975, pp. 182-205.

[20] HOUSNER, G. W., "Bending Vibrations of a Pipeline Containing Flowing Fluid", J. Appl. Mech. ASME 19, 1952, pp. 205-209.

[21] HUSEYIN, K., "Standard Forms of the Eigenvalue Problems Associated with Gyroscopic Systems", J. Sound Vib. 45(1), 1976, pp. 29-47.

[22] HUSEYIN, K., "The Effect of Damping on the Flutter Boundary of Rotating Systems", Int. Conf. Vibrations of Rotating Machinery, Cambridge, Prelim. Proc. C 180/76, I. Mech. E. England, 1976, pp. 133-138.

[23] HUSEYIN, K. and PLAUT, R. H., "Extremum Properties of the Generalized Rayleigh Quotient Associated with Flutter Instability", Quart. Appl. Math. 32, 1974, pp. 189-201.

[24] HUSEYIN, K. and PLAUT, R. H., "The Elastic Stability of Two-Parameter Nonconservative Systems", J. Appl. Mech., 40(1), 1973, pp. 175-180.

[25] McGILL, D. J., "Column Instability Under Weight and Follower Loads", J. Eng. Mech. Div. ASCE 97, 1971, pp. 629-635.

[26] FISHER, W., "Die Knicklast des einseitig eingespannten Stabes als zweiparametriges, nicht-konservatives Eigenwertproblem", ZAMM 20, 1969, pp. 544-547.

[27] NEMAT-NASSER, S. and HERRMANN, G., "Some General Considerations Concerning the Destabilising Effect in Nonconservative Systems", ZAMP 17, 1966, pp. 305-313.

PART III

by M. Zyczkowski
Technical University of Cracow
Poland

1. Influence of the Behaviour of Loading on Its Critical Value

1.1 Introduction

In most cases of structural analysis the behaviour of loading has no influence on the strength of structures. In the stability problems, however, the behaviour of loading in the course of buckling may be quite significant and may result not only in quantitative but also in qualitative effects.

For example, the critical force for a clamped column cannot be determined uniquely unless we specify the loading behaviour. Assuming for such a column the Euler force as a critical one we may commit the errors of order of hundreds per cent and even the type of the loss of stability predicted may be false.

Following the paper by Z. Kordas and M. Życzkowski (Arch.Mech.Stos. 1963/1, 7-31) we are going to analyse that problem in detail. In the literature particular attention has been paid to the case of the "tangential" force, i.e. the force which preserves its direction tangential to the deformed axis of the rod (M. Beck, [1], K. S. Deyneko, M. Y. Leonov, [14], G. Y. Dzhanelidze, [15], A. Pflüger, [13], H. Ziegler, [15]). It has been established that there occur not only significant differences as compared with the case of a force with a fixed direction - namely the critical force is here about eight times greater than the Eulerian force - but also qualitative differences, namely the critical value cannot now be determined by means of the static criterion; we have to apply here a kinetic criterion. It was proved that V. Y. Feodosyev's assertion,

[6], does not hold; the latter author found, making use of the static criterion only, that in case of the tangential loading the critical value does not exist and the rod cannot lose its stability.

First of all we introduce a classification concerning the behaviour of the force after the loss of stability. This behaviour will be described by a coefficient η which constitutes the ratio of the angle between the direction of the force (after the loss of stability of the rod) and the x-axis, to the angle of inclination of the tangent ϕ_ℓ at the free end (Fig. 1). We confine ourselves to a constant value of η during the buckling. We assume moreover that after the loss of stability the force still acts on the centre of gravity of the free cross-section (no additional moment arises) and that the deflections during the buckling are small.

We shall distinguish three fundamental ranges of variability of the coefficient η. The limiting values of η are $\eta = 0$ and $\eta = 1$; in the first case we deal with a force possessing a fixed direction (Eulerian case), while in the second - with a tangential force. In the range $\eta < 0$ we term the loading "anti-tangential", for $0 < \eta < 1$ "subtangential" and finally for $\eta > 1$ "supertangential". The tangential loading occurs for instance when the face of the rod is subjected to the action of a parallel stream of liquid or gas, the force of friction being neglected. When the influence of the friction force is taken into account then the coefficient η would diminish and thus the loading be considered as "subtangential".

The "anti-tangential" loadings are encountered frequently, e.g. in the case of rod systems (V. Y. Feodosyev, [6]; the value of the critical force can now be calculated by means of the static criterion without any appreciable difficulties. The physical interpretation of the "supertangential" loading has not so far been known to us, but the theoretical analysis of this range leads to interesting conclusions.

The coefficient η will henceforth be called the "tangency coefficient".

G. Y. Dzhanelidze in paper [5] examined the case of the subtangential loadings, $0 < \eta < 1$. He found that in the range $0 \leq \eta \leq 0.5$ the critical value of the loading can be obtained by making use of the static criterion. In the case of greater values of η, it is necessary to apply the kinetic criterion, i.e. the definition of force for which the stationary vibrations of the system pass into non-stationary. Dzhanelidze calculated the appropriate critical force in an approximate manner.

The present paper aims at determining the function $P_k = P_k(\eta)$ in the whole range of variability of η. First, we solve the problem exactly, investigating the vibrations of a system with an infinite number of degrees of freedom. It is found that for increasing η there exists an asymptotic value of the critical force which may be of a considerable theoretical importance. Next, making use of the energy method examined in [11], we derive an explicit approximate formula determining the required function $P_k = P_k(\eta)$, for the exact considerations lead to a rather complicated implicit-parametric representation of this function. The comparison with the exact results enables us to investigate in detail the accuracy of the approximate formula. In the further part of the paper we consider the influence of the distribution of the mass of the rod on its behaviour in the range of application of the kinetic criterion of stability.

1.2 Derivation of the equation determining the exact value of the critical force

Having in mind the necessity of applying the kinetic criterion of stability, we proceed to consider the vibrations of the system. The static criterion will then be obtained as a particular case, assuming that the frequency of free vibrations tends to zero.

We confine ourselves to vibrations with infinitesimal deflections (amplitudes) and angles of deflection. Then we may apply the classical equation of transverse vibrations of a beam in presence of a longitudinal force:

$$EJ\overset{\text{IV}}{y} + P\overset{\prime\prime}{y} + m\ddot{y} = 0, \tag{2.1}$$

where EJ is the rigidity of the beam, P the compressive force, m the mass per unit length. Proceeding as M. Beck, [1], and K. S. Deyneko and M. Y. Leonov, [4], we assume the following form of the solution:

$$y(x,t) = e^{i\omega t}.f(x), \tag{2.2}$$

which leads to the ordinary differential equation

$$EJf^{\text{IV}} + Pf'' - m\omega^2 f = 0. \tag{2.3}$$

Three boundary conditions corresponding to this equation concern the vanishing of the deflection, the angle of deflection and the bending moment

$$f(0) = 0, \quad f'(0) = 0, \quad f''(\ell) = 0. \tag{2.4}$$

The fourth condition refers to the transverse force at the free end of the rod. According to the definition of the transverse force and the impossibility of application of the "solidification principle" in the considered case, we calculate the transverse force as the projection of force P on the direction perpendicular to the tangent, i.e.

$$T = P(\phi_\ell - \eta\phi_\ell) = -(\eta - 1)P\phi_\ell . \tag{2.5}$$

This implies the fourth boundary condition

$$y'''(\ell) = (\eta - 1)\,\frac{P}{EJ}\,y'(\ell). \tag{2.6}$$

In the case of a tangential force ($\eta = 1$) we obviously have $y'''(\ell) = 0$.

The general solution of equation (2.3) can be written in the form

$$f(x) = A \operatorname{sh} k_1 x + B \operatorname{ch} k_1 x + C \sin k_2 x + D \cos k_2 x, \tag{2.7}$$

where k_1 and k_2 (in fact ik_2) are the roots of the corresponding characteristic equation

$$k^4 + \frac{P}{EJ} k^2 - \frac{m\omega^2}{EJ} = 0, \tag{2.8}$$

namely we have

$$k_1^2 = -\frac{P}{2EJ} + \sqrt{\frac{P^2}{4E^2J^2} + \frac{m\omega^2}{EJ}}, \quad k_2^2 = \frac{P}{2EJ} + \sqrt{\frac{P^2}{4E^2J^2} + \frac{m\omega^2}{EJ}}. \tag{2.9}$$

Taking into account the boundary conditions and setting $P/EJ = k_2^2 - k_1^2$, we arrive at the system of equations

$$\begin{aligned}
&B + D = 0\\
&Ak_1 + Ck_2 = 0,\\
&Ak_1^2 \operatorname{sh} k_1\ell + Bk_1^2 \operatorname{ch} k_1\ell - Ck_2^2 \sin k_2\ell - Dk_2^2 \cos k_2\ell = 0,\\
&Ak_1^3 \operatorname{ch} k_1\ell + Bk_1^3 \operatorname{sh} k_1\ell - Ck_2^3 \cos k_2\ell\\
&\qquad + Dk_2^3 \sin k_2\ell - (\eta - 1)(k_2^2 - k_1^2)(Ak_1 \operatorname{ch} k_1\ell\\
&\qquad + Bk_1 \operatorname{sh} k_1\ell + Ck_2 \cos k_2\ell - Dk_2 \sin k_2\ell) = 0.
\end{aligned} \tag{2.10}$$

Equating to zero the principal determinant of this system, we obtain a transcendental equation defining the frequency ω as a function of force P and the tangency coefficient η.

The results can be derived in a shorter way by eliminating by means of the first two equations constants, say, C and D and examining the determinant of second degree only. The final result can be written in the form

$$\Delta(k_1, k_2, \eta) = F_1(k_1, k_2) + \eta F_2(k_1, k_2) = 0, \tag{2.11}$$

where

$$F_1 = 2k_1^2k_2^2 - k_1k_2(k_2^2 - k_1^2)\,\mathrm{sh}\,k_1\ell \sin k_2\ell + (k_1^4 + k_2^4)\,\mathrm{ch}\,k_1\ell \cos k_2\ell,$$

$$F_2 = (k_2^2 - k_1^2)[(k_2^2 - k_1^2) + 2k_1k_2\,\mathrm{sh}\,k_1\ell \sin k_2\ell - (k_2^2 - k_1^2)\,\mathrm{ch}\,k_1\ell \cos k_2\ell]. \tag{2.12}$$

In the case of $\eta = 1$ we obtain the equation derived by M. Beck, [1], and K. S. Deyneko and M. Y. Leonov, [4].

Using equation (2.11), we can determine the critical loading both on the basis of the static and kinetic stability criteria. The static criterion is obtained by setting $k_1 = 0$ since this assumption corresponds to $\omega = 0$, i.e. the equilibrium in the "neighbouring" state. We obtain $F_1 = k_2^4 \cos k_2\ell$ and $F_2 = k_2^4(1 - \cos k_2\ell)$, and since in this case $k_2^2 = P/EJ$, the inverse function with respect to the required one is given by the formula

$$\eta = \frac{-\cos k_2\ell}{1 - \cos k_2\ell} = - \frac{\cos \pi\sqrt{p}}{1 - \cos \pi\sqrt{p}}, \tag{2.13}$$

where

$$p = \frac{P\ell^2}{\pi^2 EJ} \tag{2.14}$$

is the dimensionless compressive force.

Equation (2.13) can be inverted, namely we have

$$p = \frac{1}{\pi^2}\left(\mathrm{arc}\cos\frac{\eta}{\eta - 1}\right)^2. \tag{2.15}$$

However, for investigation, form (2.13) is more convenient. It turns out that function (2.13) has the maximum at points $p = 1, 9, 25\ \ldots, (2n - 1)^2$, their value being $\eta = 0.5$; hence function (2.15) is defined for $\eta \leq 0.5$ only. At points $p = 4, 16, 36, \ldots, (2n)^2$ the absolute value of function (2.13) tends to infinity (the function itself being negative) which corresponds to horizontal asymptotes on the graph $\eta - p$.

Equation (2.15) was in a somewhat different form given by G. Y. Dzhanelidze, [3].

Let us observe that the limiting case $\eta = 0.5$ has a simple interpretation. It corresponds to the rod with two hinges (Fig. 2), which is evident from formula (2.15) as well, since then $p = 1$.

If $\eta > 0.5$, the static criterion does not define the critical force, since the equilibrium of the rod in the "neighbouring" state cannot then exist. We apply therefore the kinetic criterion by demanding that the root of equation (2.11) determining the frequency of vibrations of the rod be a double root. Then, instead of the periodic solution (2.2), there occur functions of time of the type $te^{i\omega t}$ and the amplitude of the vibrations tends to infinity in time (this conclusion is evidently valid only in the sense of the theory of small deflections, but from the practical viewpoint it constitutes a sound stability criterion).

Root ω of equation (2.11) is double if simultaneously

$$\frac{\partial \Delta}{\partial \omega} = \left(\frac{\partial F_1}{\partial k_1}\frac{\partial k_1}{\partial \omega} + \frac{\partial F_1}{\partial k_2}\frac{\partial k_2}{\partial \omega}\right) + \eta\left(\frac{\partial F_2}{\partial k_1}\frac{\partial k_1}{\partial \omega} + \frac{\partial F_2}{\partial k_2}\frac{\partial k_2}{\partial \omega}\right) = 0. \tag{2.16}$$

Substituting the derivatives $\partial k_1/\partial\omega$ and $\partial k_2/\partial\omega$ calculated from (2.9), we obtain

$$\left(k_2\frac{\partial F_1}{\partial k_1} + k_1\frac{\partial F_1}{\partial k_2}\right) + \eta\left(k_2\frac{\partial F_2}{\partial k_1} + k_1\frac{\partial F_2}{\partial k_2}\right) = 0. \tag{2.17}$$

For a fixed value of the tangency coefficient η, equations (2.11) and (2.17) constitute a system of two transcendental equations with two unknowns k_1 and k_2. The calculation of the latter quantities leads to the required critical force and the corresponding double frequency ω, since

$$p = \frac{(k_2^2 - k_1^2)\ell^2}{\pi^2}, \qquad \omega^2 = \frac{EJ}{m}k_1^2k_2^2. \tag{2.18}$$

The solution of the system (2.11) and (2.17) for various particular values η is cumbersome. The required line $f(p,\eta) = 0$ can be found considerably easier by eliminating from these equations the variable η and reducing the system to one transcendental equation. Namely, we have

$$\eta = - \frac{F_1(k_1,k_2)}{F_2(k_1,k_2)} \tag{2.19}$$

and after substitution to (2.17)

$$k_2\left(F_2 \frac{\partial F_1}{\partial k_1} - F_1 \frac{\partial F_2}{\partial k_1}\right) + k_1\left(F_2 \frac{\partial F_1}{\partial k_2} - F_1 \frac{\partial F_2}{\partial k_2}\right) = 0. \tag{2.20}$$

Line $p = p(\eta)$ determining the critical force as a function of the tangency coefficient can now be regarded as determined in an implicit-parametric manner. Equations $(2.18)_1$ and (2.19) determine p and η as functions of two parameters k_1 and k_2, while equation (2.20) represents an implitic relation between these paramters. Solving equation (2.20) for one of the variables (for a fixed value of the other one) constitutes a much simpler procedure than solving of the system (2.11) and (2.17). However, in order to find certain characteristic points the solution of the above system is necessary.

The substitution of functions F_1 and F_2 (2.12), and their derivatives into equation (2.20) yields a comparatively long equation containing twelve basic groups of terms. Consequently numerical calculations are more conveniently carried out "gradually", by calculating separately the values of functions F_1 and F_2 and their derivatives. We present here only the formulae for these derivatives, since they will be used in our further considerations:

$$\frac{\partial F_1}{\partial k_1} = 4k_1k_2^2 + k_2(3k_1^2 - k_2^2)\ \text{sh}\ k_1\ell\ \sin k_2\ell + 4k_1^3\ \text{ch}\ k_1\ell\ \cos k_2\ell$$
$$+ (k_1^4 + k_2^4)\ell\ \text{sh}\ k_1\ell\ \cos k_2\ell + k_1k_2(k_1^2 - k_2^2)\ell\ \text{ch}\ k_1\ell\ \sin k_2\ell,$$

$$\frac{\partial F_1}{\partial k_2} = 4k_2k_1^2 + k_1(k_1^2 - 3k_2^2)\text{sh}\ k_1\ell\ \sin k_2\ell + 4k_2^3\ \text{ch}\ k_1\ell\ \cos k_2\ell$$
$$- (k_1^4 + k_2^4)\ell\ \text{ch}\ k_1\ell\ \sin k_2\ell + k_1k_2(k_1^2 - k_2^2)\ell\ \text{sh}\ k_1\ell\ \cos k_2\ell,$$

$$\frac{\partial F_2}{\partial k_1} = 4k_1(k_1^2 - k_2^2) + 2k_2(k_2^2 - 3k_1^2)\text{sh}\ k_1\ell\ \sin k_2\ell \tag{2.21}$$
$$+ 4k_1(k_2^2 - k_1^2)\ \text{ch}\ k_1\ell\ \cos k_2\ell - (k_2^2 - k_1^2)^2\ell\ \text{sh}\ k_1\ell\ \cos k_2\ell$$
$$+ 2k_1k_2(k_2^2 - k_1^2)\ell\ \text{ch}\ k_1\ell\ \sin k_2\ell,$$

$$\frac{\partial F_2}{\partial k_2} = 4k_2(k_2^2 - k_1^2) + 2k_1(3k_2^2 - k_1^2)\ \text{sh}\ k_1\ell\ \sin k_2\ell$$
$$+ 4k_2(k_1^2 - k_2^2)\ \text{ch}\ k_1\ell\ \cos k_2\ell + (k_1^2 - k_2^2)^2\ell\ \text{ch}\ k_1\ell\ \sin k_2\ell$$
$$+ 2k_1k_2(k_2^2 - k_1^2)\ell\ \text{sh}\ k_1\ell\ \cos k_2\ell.$$

We now proceed to a systematic analysis of curve $p = p(\eta)$ determined by equations (2.18), (2.19) and (2.20).

1.3 Analysis of the stability curve $p = p(\eta)$

1.3.1 Derivative $dp/d\eta$. Parameters k_1 and k_2 can be expressed again in terms of p and ω (2.9), and we may regard F_1 and F_2 (2.12) as functions of p and ω. Equation (2.20) determines then function $p = p(\omega)$ and equation (2.19) function $\eta = \eta(\omega)$, namely we have

$$\eta = -\frac{F_1[\omega, p(\omega)]}{F_2[\omega, p(\omega)]}. \tag{3.1}$$

Consider the determination of the derivative $dp/d\eta$, the knowledge of which constitutes the basis of the analysis of curve $p = p(\eta)$. In view of (3.1) we have

$$\frac{dp}{d\eta} = \frac{\frac{dp}{d\omega}}{\frac{d\eta}{d\omega}} = \frac{\frac{dp}{d\omega}F_2^2}{-\left(\frac{\partial F_1}{\partial \omega} + \frac{\partial F_1}{\partial p}\frac{\partial p}{\partial \omega}\right)F_2 + \left(\frac{\partial F_2}{\partial \omega} + \frac{\partial F_2}{\partial p}\frac{\partial p}{\partial \omega}\right)F_1}. \tag{3.2}$$

By virtue of (3.1) condition (2.16) yields

$$F_2 \frac{\partial F_1}{\partial \omega} - F_1 \frac{\partial F_2}{\partial \omega} = 0. \tag{3.3}$$

Hence, dividing by $(dp/d\omega)$, we obtain

$$\frac{dp}{d\eta} = \frac{F_2^2}{-F_2 \dfrac{\partial F_1}{\partial p} + F_1 \dfrac{\partial F_2}{\partial p}} . \tag{3.4}$$

Functions F are now treated as depending on k_1 and k_2.

Thus, in accordance with the formula

$$\frac{\partial F_1}{\partial p} = \frac{\partial F_1}{\partial k_1}\frac{\partial k_1}{\partial p} + \frac{\partial F_1}{\partial k_2}\frac{\partial k_2}{\partial p} = \frac{1}{2\sqrt{p^3 + \dfrac{4m\ell^4\omega^2}{\pi^4 EJ}}}\left(- k_1 \frac{\partial F_1}{\partial k_1} + k_2 \frac{\partial F_1}{\partial k_2}\right), \tag{3.5}$$

we obtain

$$\frac{dp}{d\eta} = \frac{2F_2^2(k_1^2 + k_2^2)\ell^2}{\pi^2\left[k_1\left(F_2 \dfrac{\partial F_1}{\partial k_1} - F_1 \dfrac{\partial F_2}{\partial k_1}\right) - k_2\left(F_2 \dfrac{\partial F_1}{\partial k_2} - F_1 \dfrac{\partial F_2}{\partial k_2}\right)\right]} . \tag{3.6}$$

Taking into account (2.20), after simplifications,

$$\frac{dp}{d\eta} = \frac{F_2^3\ell^2k_1^2}{\pi^2\left(F_2 \dfrac{\partial F_1}{\partial k_1} - F_1 \dfrac{\partial F_2}{\partial k_1}\right)} = \frac{F_2\ell^2k_1}{\pi^2\left(\dfrac{\partial F_1}{\partial k_1} + \eta \dfrac{\partial F_2}{\partial k_1}\right)} . \tag{3.7}$$

The condition of extremum (the horizontal tangent to curve $p = p(\eta)$ is therefore the following:

$$F_1 = F_2 = 0 \tag{3.8}$$

(since we must have $F_1 = 0$ in order that the value of η be finite); it is evident moreover from formulae (2.21) that condition $k_1 = 0$ does not yield extremum, since then the denominator in formula (3.7) also vanishes. The extremum condition for η (the vertical tangent) is

$$F_2 \frac{\partial F_1}{\partial k_1} - F_1 \frac{\partial F_2}{\partial k_1} = F_2 \frac{\partial F_1}{\partial k_2} - F_1 \frac{\partial F_2}{\partial k_2} = 0 \qquad (3.9)$$

and, simultaneously, $F_1 \neq 0$, $F_2 \neq 0$, $k_1 \neq 0$, and $k_2 \neq 0$.

1.3.2 The local minimum of the critical loading p. It turns out that conditions (3.8) determine the local minimum of loading p. In view of (2.12) we obtain a system of two transcendental equations for k_1 and k_2, namely

$$\begin{aligned} &2k_1^2k_2^2 - k_1k_2(k_2^2 - k_1^2)\,\mathrm{sh}\, k_1\ell \sin k_2\ell + (k_1^4 + k_2^4)\,\mathrm{ch}\, k_1\ell \cos k_2\ell = 0, \\ &(k_2^2 - k_1^2) + 2k_1k_2\,\mathrm{sh}\, k_1\ell \sin k_2\ell - (k_2^2 - k_1^2)\,\mathrm{ch}\, k_1\ell \cos k_2\ell = 0. \end{aligned} \qquad (3.10)$$

This system can be reduced to a much simpler form. We regard it at first as the system with the unknowns

$$\mathrm{sh}\, k_1\ell \sin k_2\ell = s, \qquad \mathrm{ch}\, k_1\ell \cos k_2\ell = t; \qquad (3.11)$$

the solution for these unknowns yields a much simpler system namely

$$s = \mathrm{sh}\, k_1\ell \sin k_2\ell = -\frac{k_2^2 - k_1^2}{k_1k_2}, \quad t = \mathrm{ch}\, k_1\ell \cos k_2\ell = -1. \qquad (3.12)$$

Squaring the second equation and introducing in turn sine and hyperbolic sine, we arrive at the relations

$$\mathrm{sh}\, k_1\ell = \mathrm{tg}\, k_2\ell, \quad \sin k_2\ell = -\mathrm{th}\, k_1\ell. \qquad (3.13)$$

Dividing the equations (3.12) side by side, obtain

$$\mathrm{th}\, k_1\ell\, \mathrm{tg}\, k_2\ell = \frac{k_2^2 - k_1^2}{k_1k_2}. \qquad (3.14)$$

After eliminating the tangent by means of formula (3.13) we obtain a quadratic equation for $\mathrm{ch}\, k_1\ell$ with a positive root

$$\mathrm{ch}\, k_1\ell = k_2/k_1. \qquad (3.15)$$

Consequently

$$\begin{aligned} \operatorname{sh} k_1\ell &= \operatorname{tg} k_2\ell = \frac{\sqrt{k_2^2 - k_1^2}}{k_1}, \\ \cos k_2\ell &= -\frac{k_1}{k_2}, \\ \sin k_2\ell &= -\operatorname{th} k_1\ell = -\frac{\sqrt{k_2^2 - k_1^2}}{k_2}. \end{aligned} \tag{3.16}$$

Making use of the derived relations we now prove that the minimum of the critical loading p (i.e. the consecutive values of the critical loading p) occurs always for $\eta = 0.5$. From formula (2.19) we cannot directly calculate the value of this coefficient, since an indefiniteness is obtained for it. We make use of equation (2.17):

$$\eta = -\frac{k_2 \dfrac{\partial F_1}{\partial k_1} + k_1 \dfrac{\partial F_1}{\partial k_2}}{k_2 \dfrac{\partial F_2}{\partial k_1} + k_1 \dfrac{\partial F_2}{\partial k_2}}. \tag{3.17}$$

Substituting into this equation (2.21), (3.15) and (3.16) after rather simple transformations, we obtain $\eta = 0.5$. For $\eta > 0.5$ function $p = p(\eta)$ is therefore an increasing function and since for $\eta < 0.5$ formula (2.15) holds which also defines an increasing function, we arrive at the important conclusion that in the whole range $-\infty < \eta < \infty$ function $p = p(\eta)$ is an increasing function.

We now proceed to numerical results. Instead of solving the more difficult system of transcendental equations (3.10) it is sufficient to solve (3.12) or the system of two equations selected from (3.15) and (3.16). Obviously, there exists an infinite sequence of roots k_1 and k_2 of these equations, which define the sequence of critical forces. We confine ourselves to the determination of the first two critical forces.

The systems of the transcendental equations have in this paper been solved by the method of consecutive approximation rule, "falsi rule", generalized by Gauss to systems of equations with two unknowns (cf. A. M.

Ostrowski, [12]). The geometric interpretation of this method is the following: we intend to solve the system of two equations

$$z_1 = f_1(x,y) = 0, \qquad z_2 = f_2(x,y) = 0 \tag{3.18}$$

i.e. to find the point of intersection of plane $z = 0$ by the edge of the intersection of surfaces $z_1 = f_1(x,y)$ and $z_2 = f_2(x,y)$. We select three points $P_1(x_1,y_1)$, $P_2(x_2,y_2)$, $P_3(x_3,y_3)$ lying near the expected solution. Gauss recommends here to take for instance $x_2 = x_1$ and $y_3 = y_1$; then the three points constitute the vertices of a right triangle, which considerably simplifies the calculations. We calculate the values of functions z_1 and z_2 at points P_i and through the derived points we draw two planes. The point of intersection of plane $z = 0$ by the edge of intersection of these planes, P_4, is the next approximation of the required point.

The first solution of system (3.12) is $k_1\ell = 1.63094 = 0.51914\pi$, $k_2\ell = 4.32579 = 1.37694\pi$, and consequently from formula (2.18) we have $\dot{p} = 16.0525/\pi^2 = 1.62646$. Curve $\dot{p} = p(\eta)$ therefore has at point $\eta = 0.5$ a discontinuity, since the left limit (the static criterion) is $p = 1$ while the right one (the kinetic criterion) $p = 1.62646$. The second solution of system (3.12) is $k_1\ell = 2.24953$, $k_2\ell = 10.78556$ and hence $p = 111.266/\pi^2 = 11.4145$.

1.3.3 The horizontal asymptote. The vanishing of the expression F_2 with the simultaneous condition $F_1 \neq 0$ defines a horizontal asymptote, since then $\eta \to \infty$. The determination of the corresponding critical force is of a considerable theoretical value since this is the greatest possible value of the critical loading in the considered case.

In view of (2.20) we have to solve the system of the transcendental equations

$$F_2 = 0, \qquad k_2 \frac{\partial F_2}{\partial k_1} + k_1 \frac{\partial F_2}{\partial k_2} = 0. \tag{3.19}$$

We calculate the first pair of roots only. We obtain $k_1\ell = 1.3416$, $k_2\ell = 7.0408$ and hence $p = 4.8405$. The value

$$P_\infty = 4.8405 \frac{\pi^2 EJ}{\ell^2} = 47.774 \frac{EJ}{\ell^2}, \tag{3.20}$$

is therefore the highest possible value of the critical loading for a rod fixed at one end and compressed axially. It is nearly twenty times greater than the Eulerian loading for the considered rod and therefore the increase of stability of the system due to the supertangency of the load is very large.

1.3.4 The common points of the static and kinetic criteria. If the double root of equation (2.11) i.e. the root of equation (2.20), is $\omega = 0$ then $k_1 = 0$; the corresponding critical force can be regarded as determined from both the static and kinetic criteria. The corresponding point of plane η, p is the common point of the two branches of the "stability curve".

At first sight it seems (which is not true) that this point is determined by substituting into equation (2.20) $k_1 = 0$, i.e. from the condition

$$F_2 \frac{\partial F_1}{\partial k_1} - F_1 \frac{\partial F_2}{\partial k_1} = 0. \tag{3.21}$$

However, it is seen from formulae (2.21) that the two derivatives appearing in this formula also vanish for $k_1 = 0$ and the corresponding equation determining the required common point should be defined by the limiting process, namely.

$$\left(F_2 \frac{\partial F_1}{\partial k_2} - F_1 \frac{\partial F_2}{\partial k_2}\right)\Bigg|_{k_1=0} + \lim_{k_1 \to 0} \left[\frac{k_2}{k_1}\left(F_2 \frac{\partial F_1}{\partial k_1} - F_1 \frac{\partial F_2}{\partial k_1}\right)\right] = 0. \tag{3.22}$$

A very easy transition to the limit, after the appropriate substitutions, yields a transcendental equation for k_2:

$$k_2^2\ell^2 \cos^2 k_2\ell - 2k_2\ell \sin k_2\ell \cos k_2\ell - 3k_2\ell \sin k_2\ell + 4 \sin^2 k_2\ell = 0. \quad (3.23)$$

The first three roots $k_2\ell$ of this equation together with the corresponding values of η (2.13), and the forces p, (2.18) have been collected in Table 1.

Table 1. Common points of the static and kinetic stability criteria

$k_2\ell$	η	p
4.1315	0.3543	1.7295
7.4335	-0.6898	5.5987
10.7363	0.2041	11.6791

We now prove that the common points determined in Table 1 are the points of tangency of the curves corresponding to the static and kinetic criteria. To this end, it is sufficient to compare the two derivatives $dp/d\eta$ at this point. From formulae (2.15) and (3.7) we obtain

$$\frac{dp}{d\eta} = \frac{2k_2\ell(1 - \cos k_2\ell)^2}{\pi^2 \sin k_2\ell} \quad (3.24)$$

which completes the proof.

1.3.5 Calculation of the critical tangential force. Substituting into (2.11) and (2.17) $\eta = 1$ we arrive at a system of equations determining by means of k_1 and k_2 the critical value of the tangential force (strictly speaking the sequence of critical forces the first of which only is of practical importance). This system has the form

$$k_1^4 + k_2^4 + (k_2^2 - k_1^2)k_1k_2 \text{ sh } k_1\ell \sin k_2\ell + 2k_1^2k_2^2 \text{ ch } k_1\ell \cos k_2\ell = 0,$$

$$4k_1k_2(k_1^2 + k_2^2) + (k_2^4 - k_1^4)\text{sh } k_1\ell \sin k_2\ell + k_1k_2^2(k_2^2 - 3k_1^2)\ell \text{ ch } k_1\ell \sin k_2\ell +$$

$$+ 4k_1k_2(k_1^2 + k_2^2)\text{ch } k_1\ell \cos k_2\ell + k_1^2k_2(3k_2^2 - k_1^2)\ell \text{ sh } k_1\ell \cos k_2\ell = 0. \quad (3.25)$$

The critical value of the tangential force was calculated by M. Beck, [1], and K. S. Deyneko and M. Y. Leonov, [3]. M. Beck gives the first equation (3.25) only and without going into details of the applied method he quotes the result

$$P_k = 20.05 \frac{EJ}{\ell^2}; \quad (3.26)$$

the corresponding frequency ω is not given. K. S. Deyneko and M. Y. Leonov observe that the critical force differs little from $P_k = 2\pi^2 EJ/\ell^2$ and they expand the terms of the first equation (3.25) into power series in the vicinity of this point. In this way they obtain a quadratic equation which has a double root for

$$P_k = 2.0028 \frac{\pi^2 EJ}{\ell^2} = 19.767 \frac{EJ}{\ell^2}. \quad (3.27)$$

which has the value

$$\omega^2 = 0.993 \frac{\pi^2 EJ}{m\ell^2}. \quad (3.28)$$

G. Y. Dzhanelidze [4], and V. V. Bolotin, [2], discussing these results explain the difference by the fact that the calculations are of approximate nature, they do not however state their opinion. We shall calculate the roots of the system (3.25) with a greater accuracy in order to give the reason of for the difference.

Applying a number of times the falsi rule for the system of two equations, we arrive at the results $k_1\ell = 2.20665$, $k_2\ell = 4.99201$ which yields

$$P_k = 2.03158 \frac{\pi^2 EJ}{\ell^2} = 20.0509 \frac{EJ}{\ell^2}, \quad \omega^2 = 1.24572 \frac{\pi^4 EJ}{m\ell^4} . \qquad (3.29)$$

This confirms the result of M. Beck and contradicts the result deduced by Deyneko and Leonov; even more, comparatively large difference occurs for the value of ω. Probably, the above authors committed an error in expanding into series which are in this case rather complicated.

The second pair of the roots of system (3.25) yields

$$P_{k_2} = 12.9 \frac{\pi^2 EJ}{\ell^2} = 127 \frac{EJ}{\ell^2} . \qquad (3.30)$$

It seems however, that this value is of no practical importance.

1.3.6 The graph of function $p = p(\eta)$. The above investigated points make it possible to outline the graph of the function $p = p(\eta)$, i.e. the "stability curve". The branch corresponding to the static criterion is defined by equations (2.13) and (2.15) and it is easy to draw it. To draw the branch corresponding to the kinetic criterion we have first of all used the results of the preceding analysis. Moreover, equation (2.20) has been solved for a number of a priori assumed values of k_2, selecting by means of the ordinary falsi·rule the appropriate value of k_1.

The results are presented in Fig. 3. It does not contain the curve corresponding to the double roots ω with imaginary value (on the other side of the tangency points) since it does not correspond to any interpretation - the solution is then represented by functions of the type te^{kt}. The lower branch of the curve only is of practical importance. It should be observed that in the range $0.5 < \eta < 1$ the critical force increases from the value $p = 1.626$ to the value $p = 2.032$, i.e. by about 20%. The result $p = 2.046 =$ const obtained by G. A. Dzhanelidze for the whole range constitutes therefore a rather rough approximation which is at the expense of safety.

Figures, Part III, 1

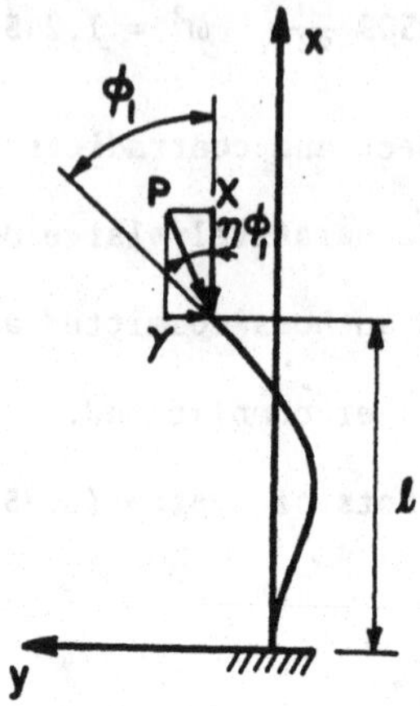

Figure 1
The rod in consideration

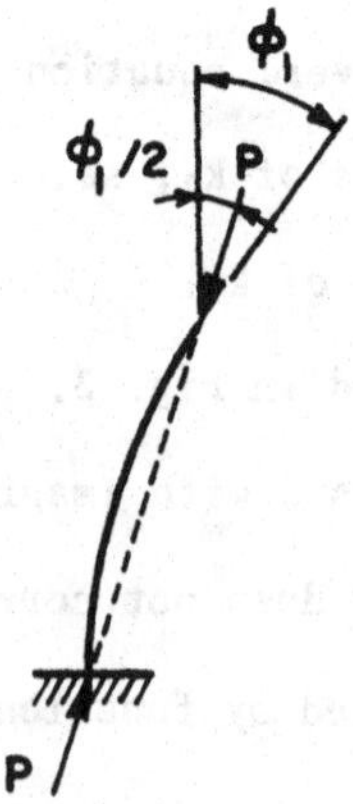

Figure 2
The case $\eta = 1/2$

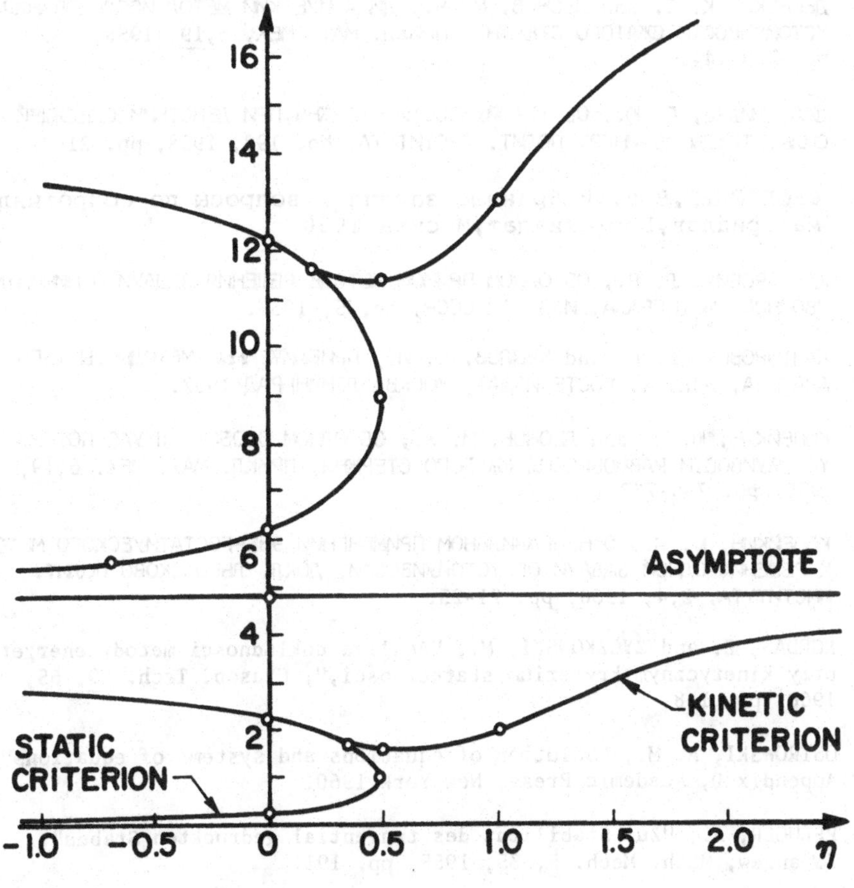

Figure 3

The stability curves $p = p(\eta)$

References, Part III, 1

[1] BECK, M., "Die Knicklast des einseitig eingespannten tengential gedrückten Stabes", ZAMP, 3,3, 1952, pp. 225.

[2] БОЛОТИН, В. В., РЕФ. ЖУРНАЛ, МЕХАНИКА, 4, 1958, No. 4535.

[3] БОЛОТИН, В. В., НЕКОНСЕРВАТИВНЫЕ ЗАДАУИ ТЕОРИИ УПРУГОИ УСТОЙУИВОСТИ, ФИЗМАТГИЗ, МОСКВА 1961.

[4] ДЕИНЕКО, К. С. and ЛЕОНОВ, М. Я., ДИНАМИУЕСКИЙ МЕТОД ИССЛЕДОВАНИЯ УСТОЙУИВОСТИ СЖАТОГО СТЕРЖНЯ, ПРИКЛ. МАТ. МЕХ, 6,19, 1955, pp. 738-744.

[5] ДЖАНЕЛИДЗЕ, Г. Ю., ОБ УСТОЙУИВОСТИ СТЕРЖНЯ ПРИ ДЕЙСТВИИ СЛЕДЯЩЕЙ СИЛЫ, ТРУДЫ ЛЕНИНГР. ПОЛИТ. ИНСТИТУТА, No. 192, 1958, pp. 21-27.

[6] ФЕОДОСЬЕВ,В.И.,Избранные задачи и вопросы по сопротивлению материалов,Гостехиздат,Москва 1950.

[7] КАНТОРОВИУ, Л. В., ОБ ОДНОМ ПРЯМОМ МЕТОДЕ РЕШЕНИЯ ЗАДАУИ О МИНИМУМЕ ДВОЙНОГО ИНТЕГРАЛА, ИЗВ. АН СССР, No. 5, 1933.

[8] КАНТОРОВИУ, Л. В. and КРЫЛОВ, В. И., ПРИБЛИЖЕННЫЕ МЕТОДЫ ВЫСШЕГО АНАЛИЗА, ИЗД, 4, ГОСТЕХИЗДАТ, МОСКВА-ЛЕНИНГРАД 1952.

[9] КОПЕЙКИН, Ю. Д. and ЛЕОНОВ. М. Я., ОБ ОДНОМ ОСОБОМ СЛУУАЕ ПОТЕРИ УСТОЙУИВОСТИ РАВНОВЕССИЯ СЖАТОГО СТЕРЖНЯ, ПРИКЛ. МАТ. МЕХ. 6,19, 1955, pp. 736-737.

[10] КОПЕЙКИН, Ю. Д., О НЕПРАВИЛЬНОМ ПРИМЕНЕНИИ ЭНЕРГОСТАТИУЕСКОГО МЕТОДА К РЕШЕНИЮ ОДНОЙ ЗАДАУИ ОБ УСТОЙУИВОСТИ, ДОКЛ. ЛЬВОВСКОВО ПОЛИТ. ННСТИТУТА, 2,4, 1960, pp. 21-25.

[11] KORDAS, Z. and ŻYCZKOWSKI, M., "Analiza dokladnosci metody energetycznej przy kinetycznym kryterium stateczności,", Czasop. Tech., 9, 65, 1960, pp. 1-8.

[12] OSTROWSKI, A. M., "Solution of equations and systems of equations", Appendix D, Academic Press, New York 1960.

[13] PFLÜGER, A., "Zur Stabilität des tangential gedrückten Stabes", Z. angew, Math. Mech. 5, 35, 1955, pp. 191.

[14] PISZCZEK, K., "Obszary rezonansowe drugiego rodzaju przy obciazeniu sledzacym", [Second kind resonance region for a load whose direction follows the deformation of the body], Rozpr. Inzyn, 2, 9, 1961, pp. 155-170.

[15] ZIEGLER, H., "Ein nichtkonservatives Stabilitätsproblem", Z. angew. Math. Mech., 8/9, 31, 1951, pp. 265.

2. Influence of Simultaneous Internal and External Damping on the Stability of Non-conservative Systems

2.1 Introduction

It is well known that within the range of applicability of kinetic criterion of stability internal damping may destabilize the system and lower the critical value of the force. Indeed, internal damping changes the type of the stability criterion: instead of looking for a double root of the characteristic equation we analyse the vanishing of the imaginary part of a complex frequency of vibrations.

The phenomenon of destabilization was first remarked by H. Ziegler [14] in 1956. Further investigations were carried out by L. M. Zoriy and M. Ya. Leonov [15], G. Herrmann and I. C. Jong [4, 5], R. C. Shieh [10]. S. Nemat-Nasser, and G. Herrmann [11] proved that the solution for a non-conservative problem without damping gives an upper bound for solutions of the corresponding problem with internal damping.

On the other hand, external damping has, as a rule, a stabilizing effect on non-conservative systems, [1]. Following the paper by A. Gajewski and M. Zyczkowski (Mechanika Teoretyczna i Stosowana 1972/1, 127-142) we give now a detailed analysis of some non-conservative systems with internal and external damping simultaneously taken into account.

2.2 Analysis of Ziegler's model

A double pendulum consisting of two rigid bars of equal length and of two hinges with elastic and viscous response is subject to a non-conservative force. The behaviour of the force is determined by the tangency coefficient η, Fig. 1. The hinges will be characterized by their elastic constants C_1 and C_2 and their viscous constants b_1 and b_2, and the masses of the bars are concentrated at the distances $\alpha\ell$ and $\gamma\ell$ from the hinges. External damping will be characterized by viscous constants $\bar{\tau}_1$ and $\bar{\tau}_2$ (the forces acting on the masses are $\bar{\tau}_1 v_1$ and $\bar{\tau}_2 v_2$ where v_1 and v_2 are the velocities).

To determine the stability of the system we consider small vibrations in the vicinity of the position of equilibrium.

Motion of the system will be described by the Lagrange equations of the second kind

$$\frac{d}{dt}\left(\frac{\partial T}{\partial \dot{\phi}_i}\right) - \frac{\partial T}{\partial \phi_i} + \frac{\partial V}{\partial \phi_i} + \frac{\partial D}{\partial \dot{\phi}_i} = Q_i, \quad i = 1,2. \tag{2.1}$$

where T denotes the kinetic energy, V - the potential of the elastic forces, D - the dissipation function, [7], Q_i - the generalized non-conservative forces. Assuming that the angles ϕ_1 and ϕ_2 are small we substitute

$$\begin{aligned}
T &= \frac{1}{2}\,\ell^2\left[\left(m_1\alpha^2 + \frac{1}{4}\,m_2\right)\dot{\phi}_1^2 + m_2\gamma\dot{\phi}_1\dot{\phi}_2 + m_2\gamma^2\dot{\phi}_2^2\right],\\
V &= \frac{1}{2}[(c_1 + c_2)\phi_1^2 - 2c_2\phi_1\phi_2 + c_2\phi_2^2],\\
D &= \frac{1}{2}[(b_1 + b_2)\dot{\phi}_1^2 - 2b_2\dot{\phi}_1\dot{\phi}_2 + b_2\dot{\phi}_2^2] +\\
&\quad + \frac{1}{2}\left[\left(\alpha^2\tau_1 + \frac{1}{4}\,\tau_2\right)\dot{\phi}_1^2 + \gamma\tau_2\dot{\phi}_1\dot{\phi}_2 + \gamma^2\tau_2\dot{\phi}_2^2\right],\\
Q_1 &= \frac{1}{2}\,P\ell(\phi_1 - \eta\phi_2), \quad Q_2 = \frac{1}{2}\,P\ell(1 - \eta)\phi_2,
\end{aligned} \tag{2.2}$$

where $\tau_1 = \ell^2\bar{\tau}_1$, $\tau_2 = \ell^2\bar{\tau}_2$. Equations (2.1) lead to the following system of linear differential equations:

$$\begin{aligned}
&\left(m_1\alpha^2 + \frac{1}{4}\,m_2\right)\ell^2\ddot{\phi}_1 + \left[(b_1 + b_2) + \left(\alpha^2\tau_1 + \frac{1}{4}\,\tau_2\right)\right]\dot{\phi}_1 + \left(c_1 + c_2 - \frac{1}{2}\,P\ell\right)\phi_1 +\\
&\quad + \frac{1}{2}\,m_2\gamma\ell^2\ddot{\phi}_2 + \left(-b_2 + \frac{1}{2}\,\gamma\tau_2\right)\dot{\phi}_2 + \left(\frac{1}{2}\,P\ell\eta - c_2\right)\phi_2 = 0,\\
&\frac{1}{2}\,m_2\gamma\ell^2\ddot{\phi}_1 + \left(-b_2 + \frac{1}{2}\,\gamma\tau_2\right)\dot{\phi}_1 - c_2\phi_1 + m_2\gamma^2\ell^2\ddot{\phi}_2 +\\
&\quad + (b_2 + \gamma^2\tau_2)\dot{\phi}_2 + \left[c_2 - \frac{1}{2}\,P\ell(1 - \eta)\right]\phi_2 = 0.
\end{aligned} \tag{2.3}$$

Assuming the solution in the form

$$\phi_1 = C_1 e^{\omega t}, \quad \phi_2 = C_2 e^{\omega t}, \tag{2.4}$$

and introducing dimensionless quantities

$$\Omega = \ell \left(\frac{m_2}{c_2}\right)^{1/2} \omega, \quad \beta = \frac{P\ell}{c_2}, \quad \mu = \frac{m_1}{m_2}, \quad \psi = \frac{c_1}{c_2},$$

$$B_i = \frac{b_i}{\ell\sqrt{m_2 c_2}}, \quad T_i = \frac{\tau_i}{\ell\sqrt{m_2 c_2}}, \quad i = 1,2 \tag{2.5}$$

we obtain the following system of linear, homogeneous equations with respect to the constants C_1 and C_2

$$\left\{\left(\mu\alpha^2 + \frac{1}{4}\right)\Omega^2 + \left[(B_1 + B_2) + \left(\alpha^2 T_1 + \frac{1}{4} T_2\right)\right]\Omega + \left(1 + \psi - \frac{1}{2}\beta\right)\right\}C_1 +$$

$$+\left\{\frac{1}{2}\gamma\Omega^2 + \left(-B_2 + \frac{1}{2}\gamma T_2\right)\Omega + \left(\frac{1}{2}\beta\eta - 1\right)\right\}C_2 = 0,$$

$$\left\{\frac{1}{2}\gamma\Omega^2 + \left(-B_2 + \frac{1}{2}\gamma T_2\right)\Omega - 1\right\}C_1 + \tag{2.6}$$

$$+\left\{\gamma^2\Omega^2 + (B_2 + \gamma^2 T_2)\Omega + \left(1 - \frac{1}{2}\beta + \frac{1}{2}\beta\eta\right)\right\}C_2 = 0.$$

Equating the determinant of this system to zero we obtain the frequency (imaginary frequency) Ω in terms of the compressive force β, the damping parameters B_i and T_i and the remaining parameters:

$$8\mu\alpha^2\gamma^2\Omega^4 + [8\gamma^2 B_1 + 2(1 + 4\gamma + 4\gamma^2 + 4\mu\alpha^2)B_2 + 8\alpha^2\gamma^2 T_1 + 8\mu\alpha^2\gamma^2 T_2]\Omega^3 +$$

$$+ \{[2(1 + 4\mu\alpha^2) + 8\gamma^2(1 + \psi) + 8\gamma] - \beta[(1 - \eta)(1 + 4\mu\alpha^2) + 4\gamma^2 + 2\eta\gamma] +$$

$$+ 8B_1B_2 + 8\gamma^2 B_1 T_2 + 2(1 + 4\gamma + 4\gamma^2)B_2T_2 + 8\alpha^2 B_2 T_1 + 8\alpha^2\gamma^2 T_1 T_2\}\Omega^2 +$$

$$+ \{4(2 - \beta + \eta\beta)B_1 + 8(\psi + \eta\beta - \beta)B_2 + 4(2\alpha^2 - \alpha^2\beta + \alpha^2\eta\beta)T_1 +$$

$$+ [2(1 + 4\gamma + 4\gamma^2) + 8\gamma^2\psi + \beta(\eta - 1 - 4\gamma^2 - 2\eta\gamma)]T_2\}\Omega +$$

$$+ [8\psi - 4(1 - \eta)(2 + \psi)\beta + 2(1 - \eta)\beta^2] = 0. \tag{2.7}$$

Motion of the system is stable if all the roots of (2.7) have their real parts negative. The Routh-Hurwitz criterion permits to determine the limits of stability (kinetic criterion, corresponding to flutter of the system). The static criterion is given by $\Omega = 0$ (buckling in the narrow sense, divergence).

To simplify further calculations we assume $\alpha = 1/4$ and $\gamma = 1/4$ (masses concentrated at the centres of the bars), and $\mu = \psi = 1$ (equal elasticity of the hinges). Equation (2.7) takes the form

$$\Omega^4 + (16B_1 + 160B_2 + T_1 + T_2)\Omega^3 + [176 - 24(2 - \eta)\beta + 256B_1B_2 + 16B_1T_2 + 144B_2T_2 + 16B_2T_1 + T_1T_2]\Omega^2 + \{128(2 - \beta + \eta\beta)B_1 + 256(1 + \eta\beta - \beta)B_2 + 8(2 + \eta\beta - \beta)T_1 + [160 + 8(2\eta - 5)\beta]T_2\}\Omega + 64[4 - 6(1 - \eta)\beta + (1 - \eta)\beta^2] = 0. \tag{2.8}$$

The static criterion $\Omega = 0$ results in vanishing free term of (2.8). It determines the critical force

$$\beta_{1,2} = \frac{3(1 - \eta) \mp \sqrt{(1 - \eta)(5 - 9\eta)}}{1 - \eta} \tag{2.9}$$

independent of the damping parameters, and is valid for $\eta < 5/9$ and $\eta > 1$.

Inside the interval $5/9 < \eta < 1$ we have to apply the kinetic criterion of stability. The Routh-Hurwitz criterion for the equation of the fourth degree

$$L\Omega^4 + M\Omega^3 + N\Omega^2 + S\Omega + R = 0 \tag{2.10}$$

has the form, [7],

$$LS^2 - MNS + M^2R < 0 \tag{2.11}$$

and together with the condition of positiveness of all coefficients of (2.10) gives sufficient conditions of stability.

Before going to detailed calculations we compare the general case with the particular case of no damping, $B_1 = B_2 = T_1 = T_2 = 0$. In the latter case we obtain $M = S = 0$, equation (2.10) turns into a biquadratic one, for which the criterion of stability limit (of a double root) has the form

$$N^2 - 4LR = 0. \tag{2.12}$$

This criterion may be obtained from (2.11) just under the assumption that M and S tend to zero in such a way as to obtain in the limit

$$\frac{M}{S} = \frac{N}{2R}, \tag{2.13}$$

In general, (2.13) is not satisfied, and then the criterion (2.11) gives for vanishing damping other results than (2.12) for zero damping.

For the system without damping we obtain from (2.12)

$$\beta_{1,2} = \frac{6(20 - 9\eta) \pm 12\sqrt{-9\eta^2 + 14\eta - 4}}{9\eta^2 - 32\eta + 32}. \tag{2.14}$$

On the other hand, taking damping into account we obtain from (2.11) the following equation, quadratic with respect to the critical force

$$A\beta^2 - B\beta + C = 0, \quad \beta_{1,2} = \frac{B \pm \sqrt{B^2 - 4AC}}{2A}, \tag{2.15}$$

where

$$\begin{aligned}
A &= b^2 - bce + 64(1 - \eta)c^2, \\
B &= 2ab - bcd - ace + 384(1 - \eta)c^2, \\
C &= a^2 - acd + 256c^2, \\
a &= 256 + 256\zeta + 16\varkappa + 160\xi\varkappa, \\
b &= (1 - \eta)(128 + 256\zeta + 8\varkappa) - 8(2\eta - 5)\xi\varkappa, \\
c &= 16 + 160\zeta + \varkappa + \xi\varkappa, \\
d &= 176 + B_1^2(256\zeta + 16\xi\varkappa + 144\zeta\xi\varkappa + 16\zeta\varkappa + \xi\varkappa^2), \\
e &= 24(2 - \eta).
\end{aligned} \tag{2.16}$$

The following damping parameters are introduced here:

$B_2/B_1 = \zeta$ - characterizes non-homogeneity of internal damping,

$T_2/T_1 = \xi$ - characterizes non-homogeneity of external damping,

$T_1/B_1 = \varkappa$ - characterizes the ratio of external to internal damping.

2.3 Particular cases

The case of no external damping is shown in Fig. 2. The parameters ξ and $\varkappa$ were substituted equal to zero, whereas $B_1 \to 0$ and $B_2 \to 0$ at constant ratio $B_2/B_1 = \zeta$. The destabilization depends significantly on the parameter ζ, and is maximal for $\zeta \to \infty$. For the tangential force, $\eta = 1$, we obtain in the limiting case $\zeta \to \infty$ the critical value ten times smaller than that calculated without damping.

In the case of no internal damping, $B_1 = B_2 = 0$, we obtain merely a slight influence of external damping on the critical force.

The case of both dampings being homogeneous ($\zeta = 1$, $\xi = 1$) is presented in Fig. 3 under the assumption that they tend to zero. Critical values depend here on the parameter $\varkappa$; destabilization is smaller than that obtained for non-homogeneous internal damping. Combining appropriately both dampings we may obtain any value of critical force between that for pure internal damping ($\varkappa = 0$) and no damping. This conclusion may explain the scatter in experimental results [6, 13].

The cases of non-homogeneous dampings are presented in Fig. 4 ($\zeta = 1$, $\xi = 5$, $\varkappa = 1/5$, internal damping homogeneous, external non-homogeneous) and in Fig. 5 ($\zeta = 5$, $\xi = 1/5$, $\varkappa = 1$, both dampings nonhomogeneous). It may be seen that non-homogeneity of damping enforces the destabilization effect.

Figures, Part III, 2

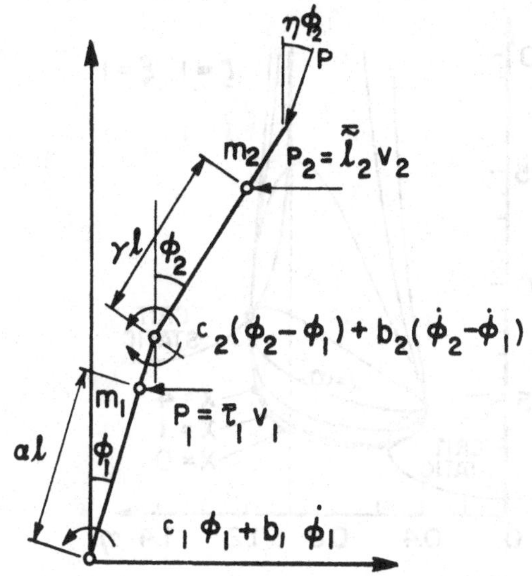

Figure 1

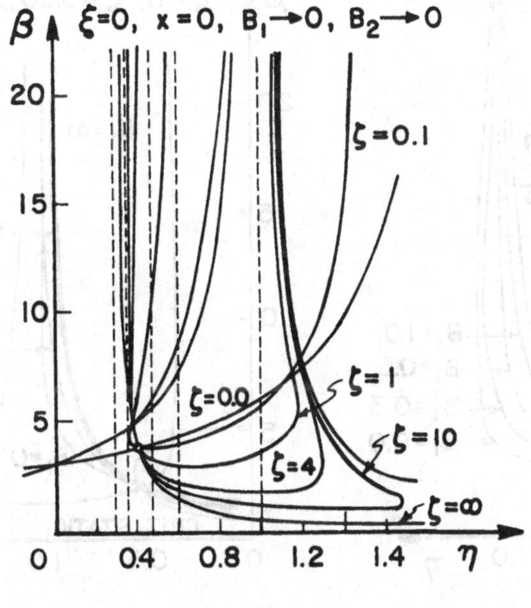

Figure 2

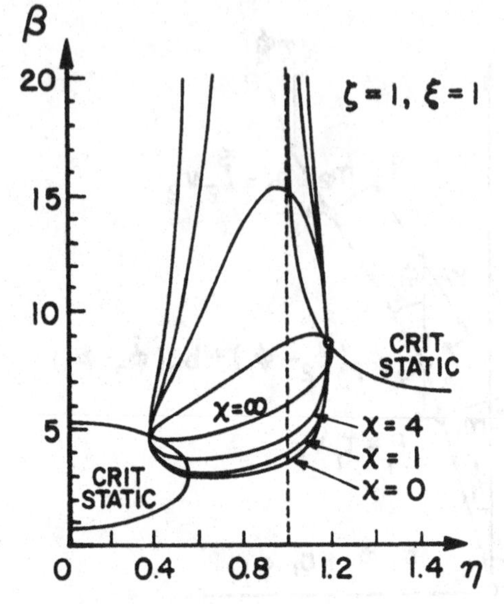

Figure 3

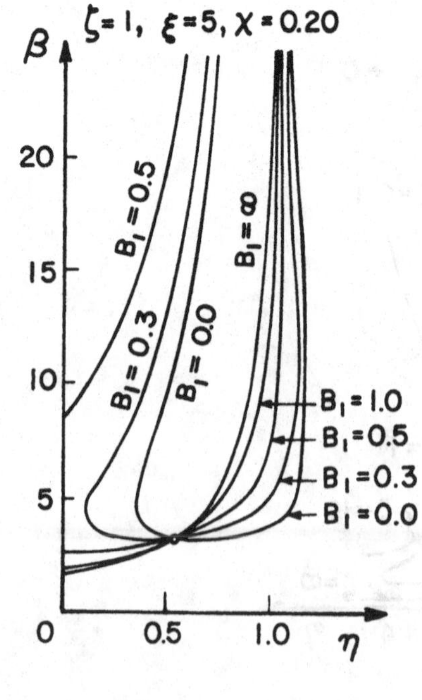

Figure 4

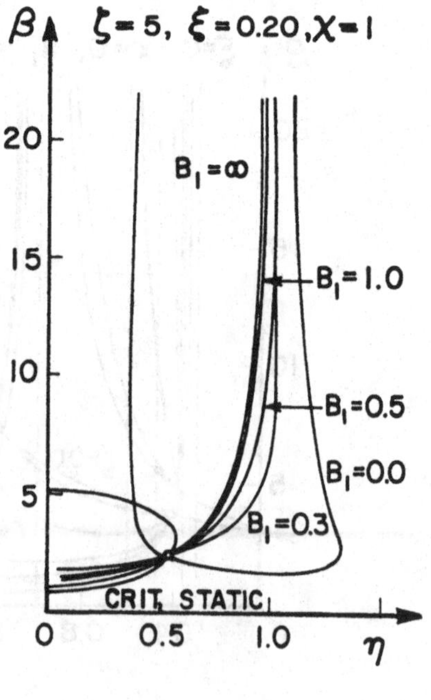

Figure 5

References, Part III, 2

[1] DŻYGADLO, Z. and SOLARZ, L., "On nonautonomous vibrations of a self-excited system with tangential force", Proc. of Vibration Problems, 2, 11, 1970, pp. 157-178.

[2] GAJEWSKI, A., "Pewne problemy optymalizacji ksztaltu pretów przy niekonserwatywnych zagadnieniach stateczności", Prace Komisji Mech. Stos. Oddz. Kraków, PAN, Mechanika Nr. 4, 1970, pp. 3-27.

[3] GAJEWSKI, A., "Badanie postaci drgán pretów ściskanych obciążeniem niekonserwatywnym", Czas. Techn. 10-M(141), 1970, pp. 1-8.

[4] HERRMAN, G. and JONG, I. C., "On the destabilizing effect of damping in nonconservative elastic systems", J. of Appl. Mech., 3, 32, 1965, pp. 592-597.

[5] HERRMANN, G. and JONG, I. C., "On nonconservative stability problems of elastic systems with slight damping", J. of Appl. Mech. 1, 33, 1966, pp. 125-133.

[6] ЯГН,Ю.И.- ПАРШИХ,Л.К.,Эхпериментальное изучение устойчивости стержня при сжатии следяшей силой, Прочн.Мат.и констр, Труды Л.П.И.,No.278, pp.52-54.

[7] KÁRMÁN, T. V. and BIOT, M. A., "Metody matematyczne w technice", PWN, Warszawa 1958.

[8] KORDAS, Z., "Stateczność preta oplywanego rownoleglym strumieniem plynu przy uwzglednieniu oporu czolowego", Rozpr. Inz., 1, 13, 1965, pp. 19-41.

[9] KORDAS, Z. and ŻYCKOWSKI, M., "Analiza dokladności metody energetycznej przy kinetycznym kryterium stateczności", Czas. Techn., 9, 35, 1960, pp. 1-8.

[10] NEMAT-NASSER, S., PRASAD, S. N. and HERRMANN, G., "Destabilizing effect of velocity-dependent forces in nonconservative continuous systems", A.I.A.A. Journal 7, 4, 1966, pp. 1276-1280.

[11] NEMAT-NASSER, S. and HERRMANN, G., "Some general consideration concerning the destabilizing effect in nonconservative systems", ZAMP, 2, 17, 1966, pp. 305-313.

[12] SHIEH, R. C., "Variational method in the stability analysis of non-conservative problems", ZAMP, 1, 21, 1970, pp. 88-100.

[13] WOOD, W. G., SAW, S. S. and SAUNDERS, P. M., "The kinetic stability of a tangentially loaded strut", Proc. Roy. Soc. Lond., A 313, 1969, pp. 239-248.

[14] ZIEGLER, H. "On the concept of elastic stability", Advances in Appl. Mech. V.4, Acad. Press, Inc., N. York 1956.

[15] ЗОРИЙ, Л. М. and ЛЕОНОВ, М. Я., ВЛИЯНИЕ ТРЕНИЯ НА УСТОЙУИВОСТЬ НЕКОНСЕРВАТИВНЫХ СИСТЕМ, ВОПРОСЫ МАШИНОСТРОЕНИЯ И ПРОУНОСТИ В МАШИНОСТРОЕНИИ, 7, 7, 1961, pp. 127-136.

3. Interaction Curves in Non-conservative Problems of Elastic Stability

3.1 Introduction

In the case of a structure loaded by several independent loadings we may present the limit of stability as a certain interaction surface in the space of those loadings. For conservative loadings it has been shown that such interaction surfaces are convex (H. Schaefer, Diss. Hannover 1934 and ZAMM 1934, 367; P. F. Papkovitch, Proc. Fourth Int. Congr. Appl. Mech. 1934, and Trudy Leningr. Korabl. Inst. 1937). It turns out, however, that for non-conservative loadings this theorem does not hold. Following the paper by Z. Kordas (Bull. Acad. Pol. Sci., 1965/5) we are going to show a simple interaction curve for a two-dimensional case which exhibits concavities both in the ranges of static and of kinetic criterion of stability.

Consider the stability problem of a bar (or plate) clamped at one end, in parallel flow of a fluid (Fig. 1). At a certain flow velocity the bar may loose rectilinear stability. Then, in addition to frontal forces (head resistance), lateral loads will occur, resulting from the inclination of the element to the direction of flow.

The existing analyzes of this problem are confined to the stability of a bar under the action of either the head resistance alone, [2], [5], [7], [8] or the lateral pressure alone, [1], [6], [9], [10]. The present paper tries to solve the problem of simultaneous action of both loads and in particular, the influence of the head resistance on the critical value of the flow velocity.

The stability problem of a bar in a fluid flow is non-conservative because the forces acting on the bar have not all a potential. Thus, the general method of analysis of critical loads should be applied, consisting in an analysis of small vibrations of the system [9].

3.2 Assumptions

It will be assumed in what follows that

a) The cross-section of the bar has, at least, one symmetry axis. We shall confine ourselves to the investigation of buckling in the symmetry plane denoting the relevant rigidity by EI.

b) We are concerned with stationary flow (the parameters characterizing the coefficients do not vary in time and the forces acting on an element of the bar depend only on the location of that element and are independent of the deformation of the bar as a whole.

c) The pressure distribution over the lateral surface will be assumed in agreement with the usual simplified flow theory.

Following V. V. Bolotin [3], we assume that the lateral pressure is

$$p = 2Bb\left(U\frac{\partial w}{\partial \xi} - \frac{\partial w}{\partial t}\right), \tag{2.1}$$

where $B = p_0 \varkappa / c_0$ is a constant characterizing the properties of the fluid. In the case of a gas, c_0 is the sound velocity, $\varkappa$ - polytropic exponent, U - flow speed, $w = w(\xi,t)$ - deflection of the bar (or plate) and b - its width.

Equation (2.1) is usually applied to problems of gas flow. In the case of a liquid, the constant B may be determined in an approximate manner by experiment.

On substituting (2.1) in the differential equation of transverse vibration of the bar we obtain:

$$EIw^{IV} + Pw'' - 2BbUw' + \rho\ddot{w} + 2Bb\dot{w} = 0, \tag{2.2}$$

where it has been assumed that $P_x = P$ (which is correct for small deformations) and the dots denote derivatives with respect to time.

3.3 Static criterion of stability

Equation (2.2) describing the vibratory motion is completed by bound conditions following from the assumption that the head resistance remains, during buckling, tangent to the deflected axis. It has, therefore, the char of a tangential force:

$$w(0,t) = 0, \quad w''(\ell,t) = 0,$$
$$w'(0,t) = 0, \quad w'''(\ell,t) = 0. \tag{3.1}$$

Assuming that the solution of equation (2.2) has the form

$$w(\xi,t) = e^{\lambda t} f(\xi), \tag{3.2}$$

we obtain, on substituting (3.2) into (2.2), the ordinary differential equation

$$EIf^{IV} + Pf'' - 2BbUf' + f\lambda(\rho\lambda + 2bB) = 0 \tag{3.3}$$

with the boundary conditions:

$$f(0) = 0, \quad f''(\ell) = 0,$$
$$f'(0) = 0, \quad f'''(\ell) = 0. \tag{3.4}$$

On setting $B = 0$ into equation (3.3) we obtain the equation studied in [8]. In the present paper the exact analysis will be based solely on the statical stability criterion alone.

This is equivalent to the research of the critical values of loads with the frequency of vibration $\omega = \lambda/i = 0$. The kinetic method (kinetic criterion) based on the coincidence of two neighbouring frequencies of natural vibration is more general. The validity range of the static and kinetic criterions will be considered in a further part of the paper using the approximate energy method.

Thus, we consider equation (3.3) for $\omega = \lambda/i = 0$. Assuming its solution in the form

$$f(\xi) = C\cdot e^{k\xi}, \tag{3.5}$$

we obtain the characteristic equation

$$k^4 + \frac{P}{EI}k^2 - \frac{2bBu}{EI}k = 0. \tag{3.6}$$

We have $k_1 = 0$ and in view of:

$$\Delta = \left(-\frac{bBU}{EI}\right)^2 + \left(\frac{P}{3EI}\right)^3 > 0. \tag{3.7}$$

$k_2 = -2\alpha$, $k_3 = \alpha + i\beta$, $k_4 = \alpha - i\beta$, where

$$\alpha = -\frac{1}{2}\left(\sqrt[3]{\frac{bBU}{EI} + \sqrt{\Delta}} + \sqrt[3]{\frac{bBU}{EI} - \sqrt{\Delta}}\right),$$

$$\beta = \frac{1}{2}\sqrt{3}\left(\sqrt[3]{\frac{bBU}{EI} + \sqrt{\Delta}} - \sqrt[3]{\frac{bBU}{EI} - \sqrt{\Delta}}\right).$$

The general integral of the equation considered is (in view of the assumption (3.5))

$$f(\xi) = C_1 + C_2\cdot e^{k_2\xi} + C_3\cdot e^{\alpha\xi}\sin\beta\xi + C_4\cdot e^{\alpha\xi}\cos\beta\xi. \tag{3.9}$$

Making use of the conditions (3.4), we obtain a set of homogeneous equa^ions in respect to constants C_i. The condition of the principal determinant becoming zero leads to a transcendental equation determining the critical value of the head resistance and the flow velocity (and, therefore, the lateral pressure) in the validity range of the static criterion of static stability. This equation has the form

$$k_2 e^{\ell(k_2 - \alpha)}\{[\alpha k_2 - (\alpha^2 - \beta^2)]\sin\beta\ell + \beta(k_2 - 2\alpha)\cos\beta\ell\} + \beta(\alpha^2 + \beta^2) = 0. \tag{3.10}$$

Equation (3.10) is an equation of the limit curve, the coordinates being the force P and the flow velocity U (α and β are expressed in terms of P and U by equations (3.8)). It will be made use of in further part of the work, for testing the accuracy of the results obtained by the energy method.

In the case of P = 0 equation (3.10) takes the form

$$2e^{3/2 k_2 \ell} \cos\left(-\frac{\sqrt{3}}{2} k_2 \ell\right) + 1 = 0. \tag{3.11}$$

This is an equation identical with that obtained by A. A. Movchan [10] by a somewhat longer and more tedious way. The solution of equation (3.11) gives the result $k_2\ell = 1.8499$ which yields the critical value of the flow velocity

$$U_{cr} = 3.1651 \frac{EI}{bB\ell^3} . \tag{3.12}$$

This result is somewhat more accurate than the results of Movchan [10] and Bolotin [3].

3.4 Energy method

Applying the energy method in the case of no potential, the motion of the system will be considered: The deflection curve will be treated as a function of not only the position but also the time and the minimum condition of potential energy will be replaced by Lagrange's equations of the second kind.

To facilitate the calculations, the origin will be assumed in this part of the paper at the free end of the undeformed bar, the senses of the axes being as in Fig. 1. We introduce the dimensionless coordinates

$$x = \frac{\xi}{\ell}, \quad y = \frac{w}{\ell} . \tag{4.1}$$

For further considerations two-parameter equation of the elastic line will be assumed in the form [7].

$$y(x,t) = q_1(t)f_1(x) + q_2(t)f_2(x), \tag{4.2}$$

where $q_1(t)$ and $q_2(t)$ are functions of time and $f_1(x)$ and $f_2(x)$ - functions of location. It will be assumed in a manner more general than before, that the head resistance P follows the tangent at the end of the bar with a certain deviation characterized by the parameter η (Fig. 2).

Equation (4.2) will satisfy the following boundary conditions:

$$\begin{aligned} &w(\ell,t) = y(\ell,t) = 0, \quad w''(0,t) = y''(0,t) = 0, \\ &w'(\ell,t) = y'(\ell,t) = 0, \quad w'''(0,t) = y'''(0,t) = 0, \end{aligned} \tag{4.3}$$

therefore it satisfies exactly the boundary conditions only for $\eta = 1$. The resulting error is small in the region $0 < \eta < 1$ but may be of essential importance for $\eta > 1$ (cf.[8]).

The Lagrange equations for the case of simultaneous lateral pressure and head resistance lead to a set of ordinary differential equations describing in the approximate manner the vibrating motion of the bar. The stability analysis of that motion determines the static stability of the bar. These equations have been derived in detail in [9]. In the present paper we quote only their final form

$$\begin{cases} A_1\ddot{q}_1 + B_1\ddot{q}_2 + C_1\dot{q}_1 + D\dot{q}_2 + E_1q_1 + F_1q_2 = 0, \\ A_2\ddot{q}_2 + B_1\ddot{q}_1 + C_2\dot{q}_2 + D\dot{q}_1 + E_2q_2 + F_2q_1 = 0, \end{cases} \tag{4.4}$$

where

$$\begin{cases} A_1 = \rho\ell^4 a_{11}, & A_2 = \rho\ell^4 a_{22}, \\ B_1 = \rho\ell^4 a_{12}, & B_1 = \rho\ell^4 a_{12}, \\ C_1 = 2bB\ell^4 a_{11}, & C_2 = 2bB\ell^4 a_{22}, \\ D = 2bB\ell^4 a_{12}, & D = 2bB\ell^4 a_{12}, \\ E_1 = E\ell b_{11} - P\ell^2 c_{11} - P\ell^2 d_{11} + & E_2 = E\ell b_{22} - P\ell^2 c_{22} - P\ell^2 d_{22} + \\ \quad + 2bB\ell^3 Ue_{11}, & \quad + 2bB\ell^3 Ue_{22}, \\ F_1 = E\ell b_{12} - P\ell^2 c_{12} - P\ell^2 d_{12} + & F_2 = E\ell b_{12} - P\ell^2 c_{12} - P\ell^2 d_{21} + \\ \quad + 2Bb\ell^3 Ue_{12}, & \quad + 2bB\ell^3 Ue_{21} \end{cases} \tag{4.5}$$

and

$$\begin{cases} a_{ij} = \int_0^1 f_i(x) f_j(x) dx, \\ b_{ij} = \int_0^1 f_i''(x) f_j''(x) dx, \\ c_{ij} = \int_0^1 f_i'(x) f_j'(x) dx, \\ d_{ij} = f_i(0) f_j'(0), \\ e_{ij} = \int_0^1 f_i(x) f_j'(x) dx. \end{cases} \tag{4.6}$$

We shall now be concerned with the analysis of the set of equations (4.4). It will be assumed that the generalized coordinates $q_1(t)$ and $q_2(t)$ are the following functions of time

$$q_1(t) = a_1 e^{kt}, \quad q_2(t) = a_2 e^{kt}, \tag{4.7}$$

where a_1 and a_2 are indeterminate constants. On substituting relations (4.7) into (4.4), we obtain a set of equations homogeneous with respect to the constants a_1 and a_2. Putting the principal determinant of this set of equations equal to zero yields an equation of the fourth order in k

$$Lk^4 + Mk^3 + Nk^2 + Sk + R = 0, \tag{4.8}$$

where

$$L = A_1A_2 - B_1^2; \; M = A_1C_2 + C_1A_2 - 2B_1D; \; R = E_1E_2 - F_1F_2; \tag{4.9}$$

$$N = A_1E_2 + E_1A_2 + C_1C_2 - B_1F_2 - B_1F_1 - D^2; \; S = C_1E_2 + E_1C_2 - DF_1 - DF_2.$$

A motion of the type $q = ae^{kt}$ is stable if the exponent k has no positive real part. The system will be at the stability limit (the Routh-Hurwitz criterion) if

$$LS^2 - MNS + M^2R = 0. \tag{4.10}$$

This is the kinetic stability criterion. We shall use it below for the determination of the equation of the limit curve (separating the stability and instability region). The static criterion will be obtained by assuming that the frequency of vibration is $\omega = k/i = 0$ (the bar is at neutral equilibrium). This leads to the equation

$$R = E_1E_2 - F_1F_2 = 0. \tag{4.11}$$

Let us observe that if the pressure p is expressed by the equation

$$p = 2BbU \frac{\partial w}{\partial \xi} = 2BUb \frac{\partial y}{\partial x} \tag{4.12}$$

(this is equivalent to the rejection of the velocity of transport, [5], [9]) the set of equations (4.4) reduces to a single differential equation of the fourth order in the form

$$L\ddot{\ddot{q}}_1 + N\ddot{q}_1 + Rq_1 = 0. \tag{4.13}$$

In this case the motion is no more stable if $R = 0$ (the static criterion), or if

$$N^2 - 4RL = 0. \tag{4.14}$$

This is the kinetic criterion for equation (4.13). The limit curve corresponds partially to the static and partially to the kinetic criterion of static stability (equations (4.11) and (4.10) or (4.14)).

Detailed calculation will be done for the following functions [7]

$$f_1(x) = x^4 - 4x + 3; \quad f_2(x) = x^5 - 5x + 5. \tag{4.15}$$

Making use of the relations (4.6), (4.5) and (4.9) we can write out the kinetic criterion (4.10) and the static criterion (4.11) thus obtaining the equations of the sought for limit curve in the form

$$13.026319E^2I^2 + 0.04110345P^2\ell^4 + 0.02286945P^2\eta^2\ell^4 - \quad (4.16)$$
$$- 0.05745925P^2\eta\ell^4 - 0.002843954B^2b^2\ell^6U^2 - 1.2653824EIP\ell^2 +$$
$$+ 0.5226024EIP\eta\ell^2 + 0.0304164EIB\ell^3U + 0.00930556P\ell^5BbU -$$
$$- 0.003317385P\ell^5bBU + 0.2171378 \frac{B^2b^2\ell^4}{\rho} EI - 0.01387642 \times$$
$$\times \frac{B^2b^2\ell^4}{\rho} P\ell^2 + 0.008672272 \frac{B^2b^2\ell^4}{\rho} P\ell^2 - 0.003045226 \frac{B^2b^2\ell^4}{\rho} Bb\ell^3U = 0;$$

$$45.71428E^2I^2 + 0.6071428P^2\ell^4 + 0.1975308b^2B^2\ell^6U^2 - \quad (4.17)$$
$$- 19.75510EIP\ell^2 + 21.71428EI\eta P\ell^2 - 15.08571EIbB\ell^3U -$$
$$- 0.5476191P^2\eta\ell^4 + 0.5714286P\ell^2bB\ell^3U - 0.4444444P\eta\ell^2B\ell^3Ub = 0.$$

If $\eta = 1$ and $U = 0$, the critical value of the tangential force is computed from equation (4.16) thus obtaining $P_k = 21.647\ EI/\ell^2$. If $P = 0$, the critical value of the flow velocity can be obtained from the static criterion (4.17). It is equal to $U_k = 3.16115\ EI/bB\ell^3$, and differs very little from (3.12).

Next, on introducing the dimensionless coordinates s and v as determined by the equations

$$P = 21.647 \frac{EI}{\ell^2} s; \quad U = 3.16115 \frac{EI}{bB\ell^3} v, \quad (4.18)$$

and the parameter $\theta = B^2b^2\ell^4/\rho EI = p_0^2\varkappa\ell/\rho EIC_0^2$, describing the influence of the velocity of transport on the form of the limit curve [9] we obtain an equation corresponding to the kinetic criterion in the form

$$v^2 - (22.4064 - 7.98778\eta)vs - (677.736 - 947.420\eta + \quad (4.19)$$
$$+ 377.084\eta^2)s^2 - (3.38330 - 0.338729\theta)v + (963.843 - 398.067\eta +$$
$$+ 10.56610 - 6.605690\eta)s - (458.362 + 7.640520) = 0,$$

and an equation corresponding to the static stability criterion in the form

$$v^2 + (19.8097 - 15.4076\eta)vs + (144.132 - 130.001\eta)s^2 - \qquad (4.20)$$
$$- 24.1593v - (216.646 - 238.132\eta)s + 23.1593 = 0.$$

Below, we shall be concerned with an analysis of the limit curves in the system s - v in the case in which the head resistance gives a tangential force ($\eta = 1$). We shall also find regions where the static and kinetic stability criterions hold. For $\eta = 1$ and for a fixed value of θ - the diagram of the function (4.19) is a hyperbola and of the function (4.20) - an ellipse. Fig. 3 gives the curves obtained for $\theta = 0$.

The ellipse obtained intersects twice the v-axis, at the points v = 1 and v = 23.159. This corresponds to stability loss under the action of the lateral pressure alone (s = 0). The two branches of the hyperbola intersect the v-axis at the points v = 23.168 and v = -19.784. In the case of v = 0 the static criterion yields no solution (the ellipse does not intersect the s-axis) but the kinetic criterion curve (the left-hand branch of the hyperbola) intersects the s-axis at the points s = 1 and s = 4.268. The value of s = 1 corresponds to the approximate critical value of the head resistance of tangential type and s = 4.268 to the other value of the critical force.

It is found next that the curves obtained (the ellipse and the hyperbola) are tangent to each other. The research of the tangency point reduces therefore to the obtainment of the coordinates of the point satisfying the condition (N = 0) and one of the equations of the limit curve.

The numbers I - VIII (Fig. 3) denote the regions and the boundaries of the regions into which the entire sv-plane is divided by the limit curves obtained. The region I is the only stability region. We are particularly interested in positive values of the drag P and the flow velocity U, i.e., with the first quadrant of the sv-plane. As is seen from Fig. 3, the stability region penetrates for $v > 0$ and $s > 0$ in the form of a sharp wedge into the sv-plane. This means that even with very high flow velocities v the motion of the bar can be stable provided that the tangential head resistance, the action of which is stabilizing, is sufficiently large.

In order to verify the energy method we shall now make use of equation (3.10) by solving it by the regula falsi rule. It is found that it is satisfied, among other, by the following pairs of roots: $s = 0$, $v = 1.0012$; $s = 0.3375$, $v = 1.5$; $s = 0.8551$, $v = 3$. The coordinates of the corresponding points obtained have here the values $s = 0$, $v = 1$; $s = 0.3306$, $v = 1.5$; $s = 0.8605$, $v = 3$. Since the points formed by means of the energy method approach closely the corresponding points determined from the accurate equation - we can infer that the energy method gives sufficiently accurate results.

Fig. 4 presents the most interesting part of the interaction curve (positive values of v ans s) for various values of the parameter θ. The concavities of this curve are clearly seen, and hence the Schaefer-Papkovitch theorem is no longer valid here.

Figures, Part III, 3

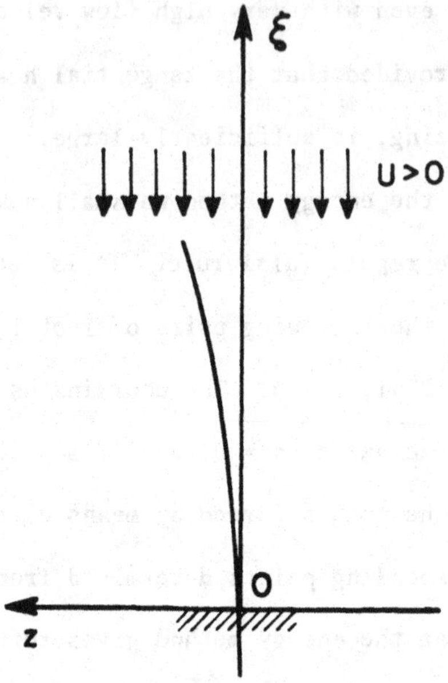

Figure 1
Scheme of the bar

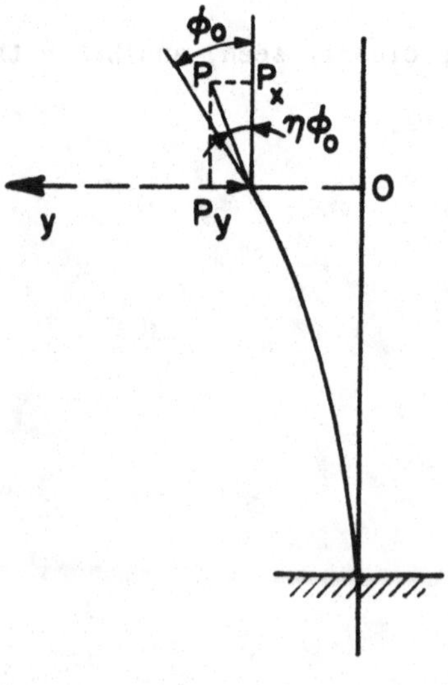

Figure 2
New set of the origin

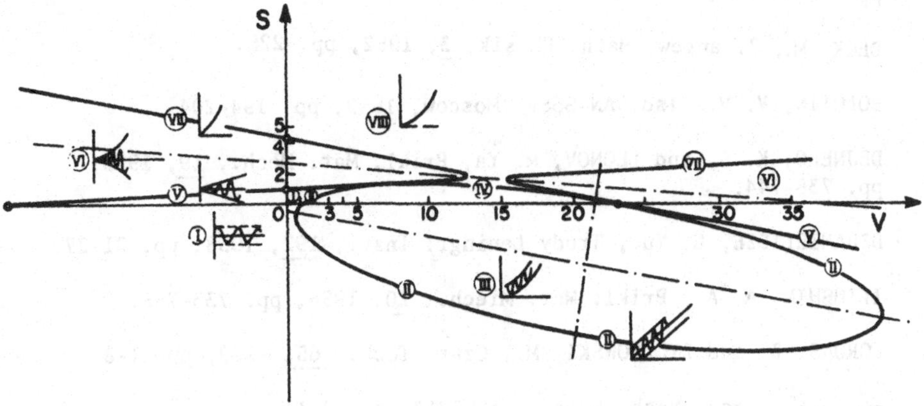

Figure 3
Limit curves $\eta = 1$, $\theta = 0$

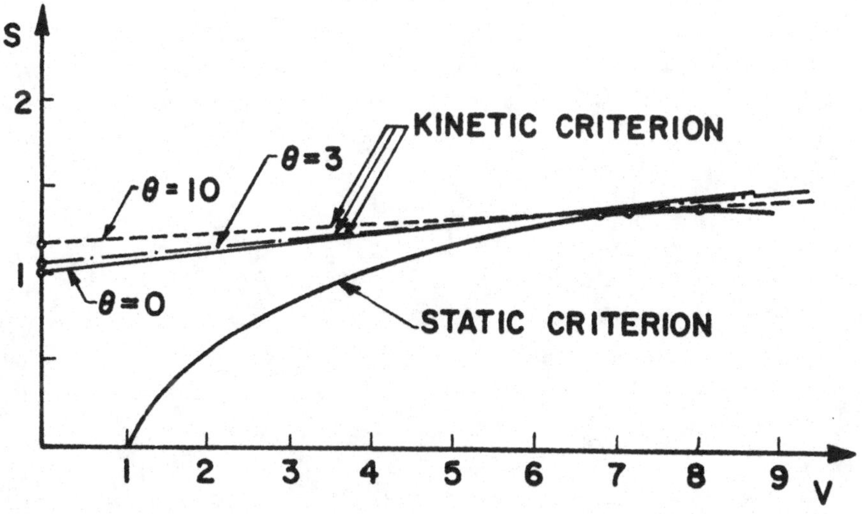

Figure 4

References, Part III, 3

[1] ASHLEY, H. and ZARTARIAN, C., J. Aeronaut. Sci., 23, 1956, pp. 1109-1118.

[2] BECK, M., Z. angew, Math. Physik, 3, 1952, pp. 225.

[3] BOLOTIN, V. V., Isd. AN SSSR, Moscow, 1959, pp. 194-204.

[4] DEJNEKO, K. S. and LEONOV, M. Ya, Prikl. Mat. Mech., 19, 1955, pp. 738-744.

[5] DZHANIELIDZE, G. Yu., Trudy Leningr. Inst., 192, 1958, pp. 21-27.

[6] ILIUSHIN A A Prikl. Mat. Miech., 20, 1956, pp. 733-755.

[7] KORDAS, Z. and ZYCZKOWSKI, M., Czas. Tech., 65, 1959, pp. 1-8.

[8] ——— , Arch. Mech. Stos., 15, 1963, pp. 7-11.

[9] KORDAS, Z., Rozpr. Inz., 1965.

[10] MOVCHAN, A. A., Prikl. Math. Miech., 20, 1956, pp. 211-222.

4. Optimal Design of Elastic Columns Subject to the General Conservative and Non-conservative Behaviour of Loading

4.1 Introduction

The behaviour of the force in the course of buckling has not only an essential influence on the critical value of that force, but also influences optimal shapes of the columns. Many papers are devoted to optimization of elastic columns compressed by a force with constant direction (T. Clausen [1], H. Blasius [2], E. L. Nikolai [3], N. G. Tchentsov [4], I. Tadjbakhsh and J. B. Keller [5]). Following the papers by A. Gajewski and M. Zyczkowski (Zeitschr. angew. Math. Physik 1970/5, 806-818; Proc. IUTAM Symp. on Instability 1969, publ. Springer 1971, 295-301) we present here the problem of optimization of compressed columns in the case of general conservative behaviour of loading and in a certain non-conservative case (anti-tangential force). Minimum weight (volume) of the bar will be assumed as the design objective. Longitudinal non-homogeneity of the material will be taken into account as well.

4.2 Statement of the problem in the general conservative case

Consider an elastically clamped non-prismatic bar of the length ℓ, compressed by a concentrated force P acting at the free end, with changing direction and point of application in the course of buckling, Figure 1. Assume, according to the paper by Z. Kordas [12], that the relative displacement of the force and its inclination with respect to the tangent at the free end are analytical functions of the deflection $w(\ell)$ and the slope $w'(\ell)$. Restriction to the analysis of infinitesimal deflections results, finally, in linear expressions for the bending moment and the shear force at the free end of the bar in terms of $w(\ell)$ and $w'(\ell)$. To obtain a more general solution we assume elastic clamping of the bar.

Introduce dimensionless cross-sectional area $\phi(\xi)$ and moment of inertia $g(\xi)$ by the formulae

$$F(\xi) = F_0\phi(\xi), \quad J(\xi) = J_0 g(\xi), \quad F_0 = F(\xi_0), \quad J_0 = J(\xi_0), \tag{2.1}$$

where ξ denotes the variable measured along the axis of the bar, $0 \leq \xi \leq \ell$, F and J denote cross-sectional area and moment of inertia, respectively, and ξ_0 refers to a suitable chosen point from the interval $0 \leq \xi \leq \ell$. A relatively broad class of non-prismatic bars may be described by the relation

$$\phi(\xi) = [g(\xi)]^{\varkappa}, \tag{2.2}$$

where $\varkappa$ = const. The value $\varkappa = 1/2$ describes similar cross-sections (spatially tapered bar); $\varkappa = 1$ or $\varkappa = 1/3$ correspond to affine cross-sections (plane-tapered bar, buckling perpendicular to the plane of tapering and buckling in the plane of tapering, respectively), Figure 2. The eventual longitudinal elastic non-homogeneity of the bar will be described by the function $f(\xi)$,

$$E = E_0 f(\xi), \tag{2.3}$$

where $E_0 = E(\xi_0)$.

The differential equation of the deflection line of the considered non-prismatic column may be written in the form

$$\frac{d^2}{d\xi^2}\left(EJ\,\frac{d^2w}{d\xi^2}\right) + P\,\frac{d^2w}{d\xi^2} = 0, \tag{2.4}$$

where w stands for the deflection. Introducing dimensionless variables x, y and dimensionless force β by the formulae

$$x = \frac{\xi}{\ell}, \quad y = \frac{w}{\ell}, \quad \beta = \frac{P\ell^2}{E_0 J_0} \tag{2.5}$$

we rewrite (2.4) in the form

$$[f(x)g(x)y'']'' + \beta y'' = 0; \tag{2.6}$$

dashes denote differentiation with respect to the variable x.

The boundary conditions, corresponding to the considered case of loading take the form

$$y(0) = 0, \quad y'(0) = \frac{1}{\beta}\zeta(fgy'')_{x=0}, \quad (fgy'')_{x=1} = -\beta[\rho y'(1) + \vartheta y(1)],$$

$$[(fgy'')']_{x=1} = -\beta[\mu y'(1) + \nu y(1)], \tag{2.7}$$

where the parameter ζ characterizes the elasticity of clamping ($\zeta = 0$, rigidily clamped bar; $\zeta \to \infty$ for an ideal hinge), and the parameters ρ, ϑ, μ, ν, describing the behaviour of the force in the course of buckling, are to be determined in individual cases of loading (some examples will be given in sec. 4). In general, the behaviour of the force described by (2.7) is non-conservative; in what follows we confine ourselves to the conservative case. To find the condition of conservativeness we express the moment M and the horizontal component of the force H in terms of the slope $\alpha = w'(\ell)$ and deflection $a = w(\ell)$ as follows (making use of (2.7)):

$$M = -\beta(\rho\alpha + \vartheta a), \quad H = \beta(\mu\alpha + \nu a) - \beta\alpha. \tag{2.8}$$

Considering infinitesimal deflections only, we assume the vertical component of the force P to be constant, hence conservative. The condition of existence of the potential

$$\frac{\partial H}{\partial \alpha} = \frac{\partial M}{\partial a} \tag{2.9}$$

leads to the following relation between the parameters:

$$\vartheta = 1 - \mu. \tag{2.10}$$

It will be shown that the fulfilment of the condition (2.10) results in an essential simplification of the optimization. Some typical examples of conservative behaviour of loading are shown in Figure 3 with the corresponding values of the parameters ρ, ϑ, μ, and ν; the relation (2.10) is in all cases obviously satisfied.

Integrating equation (2.6) twice and introducing the new dependent variable v by the formula

$$v(x) = y(x) - \frac{B_1}{\beta} - \frac{B_2}{\beta} x \tag{2.11}$$

(where B_1 and B_2 denote the constants of integration), we arrive at the equation

$$f(x)g(x)v'' + \beta v = 0 \tag{2.12}$$

with the following two boundary conditions, resulting from (2.7) and the substitution (2.11):

$$(1 - \vartheta)v(1) - \rho v'(1) + (\rho + \vartheta)v'(0) + (\rho\zeta + \vartheta\zeta + \vartheta)v(0) = 0,$$

$$\nu v(1) + (\mu - 1)v'(1) - (\mu + \nu)v'(0) - (\mu\zeta + \nu\zeta + \nu)v(0) = 0. \tag{2.13}$$

The problem of optimalization will be now stated as follows. We look for such a function $\phi(x)$, which yields the minimal volume of the column.

$$V = F_0 \ell \int_0^1 \phi(x)dx = \min \tag{2.14}$$

under the constraint of the given critical force, determined by the differential equation (2.12) and the boundary conditions (2.13). Following Tchentsov's suggestion [4], we reduce this Lagrangean problem to an ordinary problem of variational calculus, simply eliminating the function $\phi(x)$ by means of (2.2) and (2.12)

$$V = F_0 \ell \int_0^1 \left(- \frac{\beta v}{f(x) v''} \right)^{\varkappa} dx = \min. \tag{2.15}$$

The function f(x), determining the non-homogeneity, is assumed to be known.

The Euler-Lagrange equation, corresponding to the functional (2.15) is of the fourth order and requires, in principle, four boundary conditions. The considered functional is homogeneous of degree zero with respect to the function v, and this fact reduces the number of necessary boundary conditions to three. We have, however, only two boundary conditions imposed on the function v(x), (2.12) at our disposal; thus, the problem reduces to that with variable end-points, and suitable transversality conditions will be applied.

4.3 General considerations

The functional (2.15) is of the form

$$V = \int_0^1 \tilde{F}(x, v, v'') dx. \tag{3.1}$$

Introducing, after Tchentsov, the notation

$$t = \frac{\partial \tilde{F}}{\partial v''} = F_0 \ell \varkappa \beta^{\varkappa} \frac{v^{\varkappa}}{(-v'')^{\varkappa+1} f^{\varkappa}}, \tag{3.2}$$

we obtain the Euler-Lagrange equation in the form

$$vt'' - tv'' = 0. \tag{3.3}$$

Thus, the first integration yields

$$vt' - tv' = C \tag{3.4}$$

or, with substituted t,

$$\left.\begin{aligned} & v^{\varkappa}[\varkappa(\ell n\ f)' v v'' + (1-\varkappa) v' v'' + (1+\varkappa) v v'''] = \bar{C}\ f^{\varkappa}(-v'')^{\varkappa+2} \\ & \bar{C} = - \frac{C}{F_0 \ell \varkappa \beta^{\varkappa}} \end{aligned}\right\} \tag{3.5}$$

According to our previous considerations we have now to determine the constant C from the transversality condition. In the Eulerian case of the force with constant direction Tchentsov found C = 0 considering the conditions of symmetry of the deflection line. We are going to prove that C = 0 for the general conservative behaviour of loading.

The transversality condition for the functional (3.1) may be written in the form

$$\left.\begin{aligned} &\left[-\frac{d}{dx}\left(\frac{\partial\tilde{F}}{\partial v''}\right)\right]_{x=1}\delta v(1) + \left.\frac{\partial\tilde{F}}{\partial v''}\right|_{x=1}\delta v'(1) + \left[\frac{d}{dx}\left(\frac{\partial\tilde{F}}{\partial v''}\right)\right]_{x=0}\delta v(0) \\ &\qquad - \left.\frac{\partial\tilde{F}}{\partial v''}\right|_{x=0}\delta v'(0) = 0 \end{aligned}\right\} \quad (3.6)$$

(cf. M. A. Lavrentiev, L. A. Lusternik [18]).

This condition, together with two relations derived from the boundary conditions (2.13), yields the following system of equations for the unknown variations $\delta v_0 = \delta v(0)$, $\delta v_1 = \delta v(1)$, $\delta v_0' = \delta v'(0)$ and $\delta v_1' = \delta v'(1)$:

$$\begin{aligned} &t_1'\delta v_1 - t_1\delta v_1' - t_0'\delta v_0 + t_0\delta v_0' = 0, \\ &(1 - \vartheta)\delta v_1 - \rho\delta v_1' + (\rho\zeta + \vartheta\zeta + \vartheta)\delta v_0 + (\rho + \vartheta)\delta v_0' = 0, \qquad (3.7) \\ &\nu\delta v_1 + (\mu - 1)\delta v_1' - (\mu\zeta + \nu\zeta + \nu)\delta v_0 - (\mu + \nu)\delta v_0' = 0. \end{aligned}$$

In view of that fact equation (3.5) is homogeneous of degree $\varkappa + 2$ with respect to v, and that the conditions (3.7) are homogeneous as well, one of the quantities v_0, v_0, v_1, v_1 may be arbitrarily fixed; this means that one of the variations δv_0, δv_1, $\delta v_0'$ or $\delta v_1'$ may vanish. Equating successively all the four variations to zero we arrive at four systems of three linear homogeneous algebraic equations each, which determine in turn the remaining three variations. These variations must not all be equal to zero, hence the corresponding principal determinants have to vanish, i.e.

$$\left.\begin{aligned}
&t_1'[(\rho\nu - \mu\vartheta)(\zeta + 1) + (\rho + \vartheta)\zeta + \vartheta] - t_1[(\rho\nu - \mu\vartheta)\zeta + (\mu + \nu)\zeta \\
&\qquad + \nu] - t_0[(\rho\nu - \mu\vartheta) + (\mu + \vartheta - 1)] = 0, \\
&t_1'[(\rho\nu - \mu\vartheta) + (\rho + \vartheta)] - t_1[(\rho\nu - \mu\vartheta) + (\mu + \nu)] + t_0[(\rho\nu - \mu\vartheta) \\
&\qquad + (\mu + \vartheta - 1)] = 0, \\
&t_1'(\rho\nu - \mu\vartheta) - t_0'[(\rho\nu - \mu\vartheta) + (\mu + \nu)] - t_0[(\rho\nu - \mu\vartheta)\zeta \\
&\qquad + (\mu + \nu)\zeta + \nu] = 0, \\
&t_1(\rho\nu - \mu\vartheta) - t_0[(\rho\nu - \mu\vartheta) + (\lambda + \vartheta) - t_0[(\rho\nu - \mu\vartheta)(\zeta + 1) \\
&\qquad + (\rho + \vartheta)\zeta + \vartheta] = 0.
\end{aligned}\right\} \tag{3.8}$$

It may be shown easily that only one of the conditions (3.8) is independent. In solving particular problems we have to choose a form of the condition (3.8) which is not identically satisfied.

Rewrite now (3.4) in the form

$$\left(\frac{t}{v}\right)' = \frac{C}{v^2}. \tag{3.9}$$

Integration yields here

$$t = C_1 v + Cv \int_{x_0}^{x} \frac{dx}{v^2}, \tag{3.10}$$

where C is a constant of integration, different from zero. Denote now

$$\int_{x_0}^{x} \frac{dx}{v^2} = \Phi(x), \quad \Phi(0) = \Phi_0, \quad \Phi(1) = \Phi_1; \tag{3.11}$$

the conditions (3.8) take the form

$$\left.\begin{aligned}
-v_0'(\mu + \vartheta - 1) = \frac{C}{C_1}\Big\{ &-\left(\frac{1}{v_1} + v_1'\Phi_1\right)[(\rho\nu - \mu\vartheta)(\zeta + 1) \\
&+ (\rho + \vartheta)\zeta + \vartheta] + v_1\Phi_1[(\rho\nu - \mu\vartheta) + (\mu + \nu)\zeta + \nu] \\
&+ \left(\frac{1}{v_0} + v_0'\Phi_0\right) + [(\rho\nu - \mu\vartheta) + (\mu + \vartheta - 1)]\Big\}.
\end{aligned}\right\} \tag{3.12}$$

$$\left.\begin{aligned} v_0(\mu + \vartheta - 1) = \frac{C}{C_1}\Big\{ -\left(\frac{1}{v_1} + v_1'\Phi_1\right)[(\rho\nu - \mu\vartheta) + (\rho + \vartheta] \\ + v_1\Phi_1[(\rho\nu - \mu\vartheta) + (\mu + \nu)] - v_0\Phi_0[(\rho\nu - \mu\vartheta) + (\mu + \vartheta - 1)]\Big\}, \end{aligned}\right\} \quad (3.13)$$

$$\left.\begin{aligned} -v_1'(\mu + \vartheta - 1) = \frac{C}{C_1}\Big\{ -\left(\frac{1}{v_1} + v_1'\Phi_1\right)(\rho\nu - \mu\vartheta) + \left(\frac{1}{v_0} + v_0'\Phi_0\right) \\ [(\rho\nu - \mu\vartheta) + (\mu + \nu)] + v_0\Phi_0[(\rho\nu - \mu\vartheta)\zeta + (\mu + \nu)\zeta + \nu]\Big\}, \end{aligned}\right\} \quad (3.14)$$

$$\left.\begin{aligned} -v_1(\mu + \vartheta - 1) = \frac{C}{C_1}\Big\{ - v_1\Phi_1(\rho\nu - \mu\vartheta) + \left(\frac{1}{v_0} + v_0'\Phi_0\right) \\ [(\rho\nu - \mu\vartheta) + (\rho + \vartheta)] + v_0\Phi_0[(\rho\nu - \mu\vartheta)(\zeta + 1) + (\rho + \vartheta)\zeta + \vartheta]\Big\}. \end{aligned}\right\} \quad (3.15)$$

It is seen that in the case of the conservative behaviour of the force at least one of the conditions (3.12) - (3.15) in view of (2.10) leads to the conclusion $C = 0$. Simultaneous vanishing of all coefficients of C in the equations (3.12) - (3.15) is impossible since then the volume of the bar would be independent of C. In non-conservative cases the constant C is, in general, different from zero; some particular solutions of such problems are given in [16], and will be discussed in section 5.

In the following we confine ourselves to homogeneous bars, putting $f(x) \equiv 1$ in the equation (3.5). Certain solutions for non-homogeneous bars, compressed by a polar force, are given in [13]. Equation (3.5) may be once more integrated, yielding

$$v''v^{(1-\varkappa)/(1+\varkappa)} = C_2, \tag{3.16}$$

C_2 being the constant of integration. General solution of (3.16) may be written in the form of the inverse function

$$x = \pm \int \frac{\delta v}{A_1\, v^{2\varkappa/(1+\varkappa)} + A_2} + A_3. \tag{3.17}$$

In the most important cases when $\varkappa = 1$, 1/2 and 1/3 the integration in (3.7) may be performed and the exact solution given in a closed form. The two boundary conditions (2.13) determine two of the constants in terms of the third, which may be chosen arbitrarily and does not influence the optimal shape of the bar, i.e. the function $\phi(x)$, evaluated from (2.12).

4.4 Particular solutions

4.4.1 Plane-tapered bar, $\varkappa = 1$. In this case the solution is particularly simple if the deflection line has no inflection point, i.e. if the curvature v'' does not change the sign:

$$v = A_1 + A_2 x + A_3 x^2 . \tag{4.1}$$

Assume $v(0) = 1$, then $A_1 = 1$; the boundary conditions (2.13) determine A_2 and A_3:

$$A_2 = \frac{\zeta(2\rho + 2\vartheta - \mu - \nu) - (2\mu + \nu - 2)}{2\mu + \nu - 2\rho - \vartheta - 1}, \quad A_3 = \frac{\zeta(\mu + \nu - \rho - \vartheta) - 1}{2\mu + \nu - 2\rho - \vartheta - 1} . \tag{4.2}$$

The shape of the optimal bar is given by (2.12) and the relation $g(x_0) = 1$ (x_0 stands for the coordinate of a point which, in general, may be chosen arbitrarily):

$$g(x) = \frac{1 + A_2 x + A_3 x^2}{1 + A_2 x_0 + A_3 x_0^2} \tag{4.3}$$

$$\beta = \frac{2A_3}{1 + A_2 x_0 + A_3 x_0^2} . \tag{4.4}$$

The relation (4.4) determines the critical force, or, strictly speaking, the "basic rigidity" $E_0 J_0$ for the given critical force.

Example

Consider the loading by a liquid in a container, Figure 3c. The formulae (4.2) yield here $A_2 = 0$, $A_3 = -1/(1 - 2\rho)$. Choosing $x_0 = 0$, we obtain from (4.2) and (4.4) $\beta = 2/(1 - 2\rho)$ and $g(x) = 1 - x^2/(1 - 2\rho)$. The first of these relations determines the rigidity $E\,J_0$ in terms of the force (height of the column of liquid), or height in terms of $E\,J_0$. In the latter case the unknown h appears in β as well as in ρ; we obtain a quadratic equation, whence

$$\frac{h}{\ell} = \frac{1}{8}\left\{-1 + \sqrt{1 + 32\left[\frac{E\,J_0}{\gamma\pi R^2 \ell^3} - \left(\frac{R}{\ell}\right)^2\right]}\right\}$$

Assuming for example, $\ell = 100$ cm, $J_0 = 1/6$ cm^4, $R = 25$ cm, $E = 2 \times 10^9$ G/cm^2, $\gamma = 1$ G/cm^2, we obtain $h/\ell = 0.137$ and $g(x) = 1 - 0.407\,x^2$. The shape of the optimal bar is shown in Figure 4.

4.4.2 Spatially tapered bar, $\varkappa = 1/2$. In the case the solution (3.17) may be given in a parametrical form:

$$v = A_1 \sin^3 z, \quad x = A_3 + A_2\left(z - \frac{1}{2}\sin 2z\right). \tag{4.10}$$

Assume that the value of the parameter $z = z_0$ corresponds to $x = 0$ and $z = z_1$ corresponds to $x = 1$. This assumption and the boundary conditions (2.13) give

$$\left.\begin{aligned}
A_3 &= \frac{3[\rho\cos z_1 - (\rho + \vartheta)\cos z_0]}{2[(1-\vartheta)\sin^3 z_1 + \vartheta(\rho\zeta + \vartheta\zeta + \vartheta)\sin^3 z_0]} \\
&= \frac{3[(1-\mu)\cos z_1 + (\mu + \nu)\cos z_0]}{2[\nu\sin^3 z_1 - (\mu\zeta + \nu\zeta + \nu)\sin^3 z_0]}, \\
A_3 &= -A_2\left(z_0 - \frac{1}{2}\sin 2z_0\right). \\
2[(1-\vartheta)\sin^3 z_1 &+ (\rho\zeta + \vartheta\zeta + \vartheta)\sin^2 z_0] = 3[\rho\cos z_1 - \\
&(\rho + \vartheta)\cos z_0]\left[(z_1 - z_0) - \frac{1}{2}(\sin 2z_1 - \sin 2z_0)\right], \\
2[\nu\sin^3 z_1 &- (\mu\zeta + \nu\zeta + \nu)\sin^3 z_0] = 3[(1-\mu)\cos z_1 + \\
&(\mu + \nu)\cos z_0]\left[(z_1 - z_0) - \frac{1}{2}(\sin 2z_1 - \sin 2z_0)\right],
\end{aligned}\right\} \tag{4.11}$$

whereas the constant A_1 remains arbitrary. The two last equations constitute the relations between z_0 and z_1 and the parameters determining the behaviour of loading. The shape of the optimal bar and the parameter of the critical force are given by

$$g(z) = \frac{\sin^4 z}{\sin^4 z^*}, \qquad \beta = \frac{3}{4}\frac{1}{A_2^2 \sin^4 z^*}, \tag{4.12}$$

where z^* denotes the value of the parameter corresponding to x_0 and $J = J_0(g(z^*) = 1)$.

4.4.3. Plane-tapered bar, $\varkappa = 1/3$. Similarly as in the preceding case we here present the solution in parametrical form

$$v = A_1 \sin^4 z, \qquad x = A_3 + A_2 \cos z(3 - \cos^2 z), \tag{4.13}$$

denoting by z_0 and z_1 the values of the parameter corresponding to $x = 0$ and $x = 1$, respectively. These relations and the boundary conditions (2.13) give

$$\left.\begin{aligned}
A_2 &= \frac{4[(\rho + \vartheta)\cos z_0 - \rho \cos z_1]}{3[(1 - \vartheta)\sin^4 z_1 + (\rho\zeta + \vartheta\zeta + \vartheta)\sin^4 z_0]} \\
&= \frac{4[(\mu - 1)\cos z_1 - (\mu + \nu)\cos z_0]}{3[\nu \sin^4 z_1 - (\mu\zeta + \nu\zeta + \nu)\sin^4 z_0]}, \\
A_3 &= -A_2 \cos z_0(3 - \cos^2 z_0), \\
3[(1 - \vartheta)\sin^4 z_1 &+ (\rho\zeta + \vartheta\zeta + \vartheta)\sin^4 z_0] = 4[(\rho + \vartheta)\cos z_0 - \\
&\rho \cos z_1][\cos z_1(3 - \cos^2 z_1) - \cos z_0(3 - \cos^2 z_0)], \\
3[\nu \sin^4 z_1 &- (\mu\zeta + \nu\zeta + \nu)\sin^4 z_0] = 4[(\mu - 1)\cos z_1 - \\
&(\mu + \nu)\cos z_0][\cos z_1(3 - \cos^2 z_1) - \cos z_0(3 - \cos^2 z_0)],
\end{aligned}\right\} \tag{4.14}$$

whereas the constant A remains arbitrary. The shape of the optimal bar and the parameter of the critical force are given by

$$g(z) = \frac{\sin^6 z}{\sin^6 z^*}, \qquad \beta = \frac{4}{9}\frac{1}{A_2^2 \sin^6 z^*}, \tag{4.15}$$

where the meaning of z* is as before.

4.5 Optimization of an elastic bar compressed by antitangential force

As an example of non-conservative behaviour of loading we consider the optimization of a column compressed by an antitangential force (cf. Lecture I). The formulation of the problem corresponds to any value of the tangency coefficient η, but it turns out that the static criterion of stability applied in what follows restricts the solution to negative values of η.

The derivation of the Euler-Lagrange equation (3.5) proceeds as before. The constant C, however, is here no longer equal to zero, and the transversality condition

$$\begin{aligned} &-\left[\left(F_{v'} - \frac{d}{dx}F_{v''}\right)_0 \delta v_0 + (F_{v''})_0\, \delta v_0'\right] + \\ &+\left[\left(F_{v'} - \frac{d}{dx}F_{v''}\right)_1 \delta v_1 + (F_{v''})_1\, \delta v_1'\right] = 0 \end{aligned} \tag{5.1}$$

yields in the simplest case $\varkappa = 1$ (plane tapered columns) and f = 1 (homogeneous columns)

$$\left(\frac{v}{v''^2}\right)_0 + \frac{1-\eta}{\eta}\left(\frac{v}{v''^2}\right)_1 = 0. \tag{5.2}$$

The functions v and v'' are equal to zero at the point x = 1. Using (3.5) and de l'Hospital rule we obtain finally

$$v_0' = C v_0''^2. \tag{5.3}$$

This equation determines the integration constant C. It may be satisfied only for negative values of η, and hence we restricted our considerations to antitangential forces.

In the case of $C \neq 0$ equation (3.5) cannot be integrated exactly in closed form. A computer solution is presented in Fig. 5: it shows the profit in material Z as a function of the tangency coefficient η. The solution of (3.5) for $C = 0$ is also shown for comparison: it is much worse than the solution making use of the condition (5.3).

Figures, Part III, 4

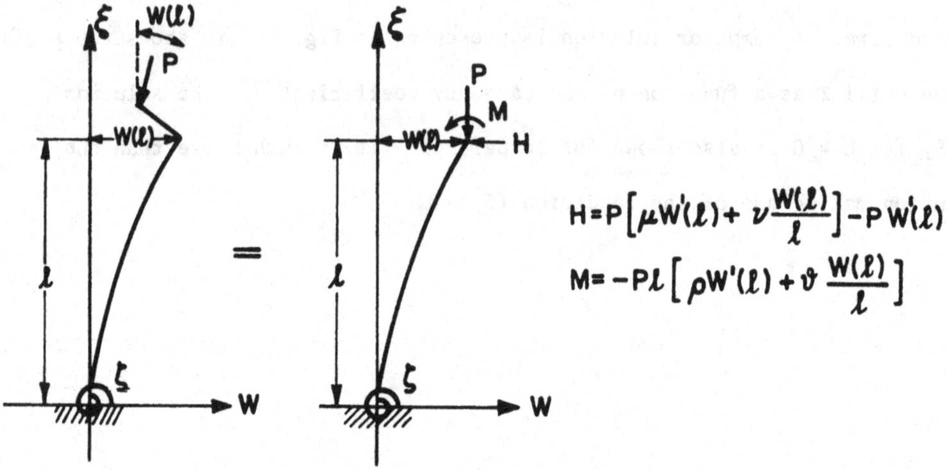

Figure 1

Scheme of the bar

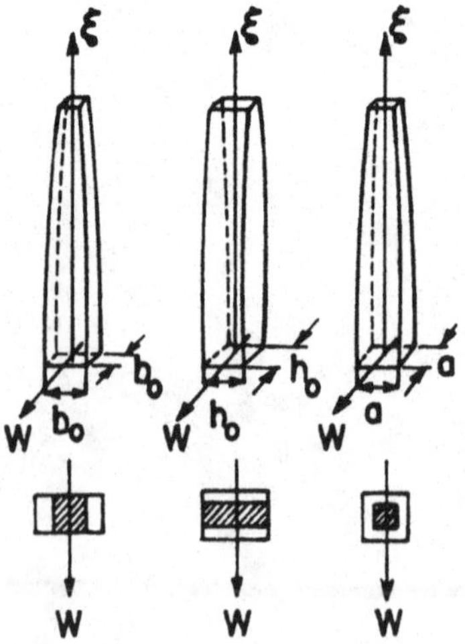

Figure 2

Typical tapered bars

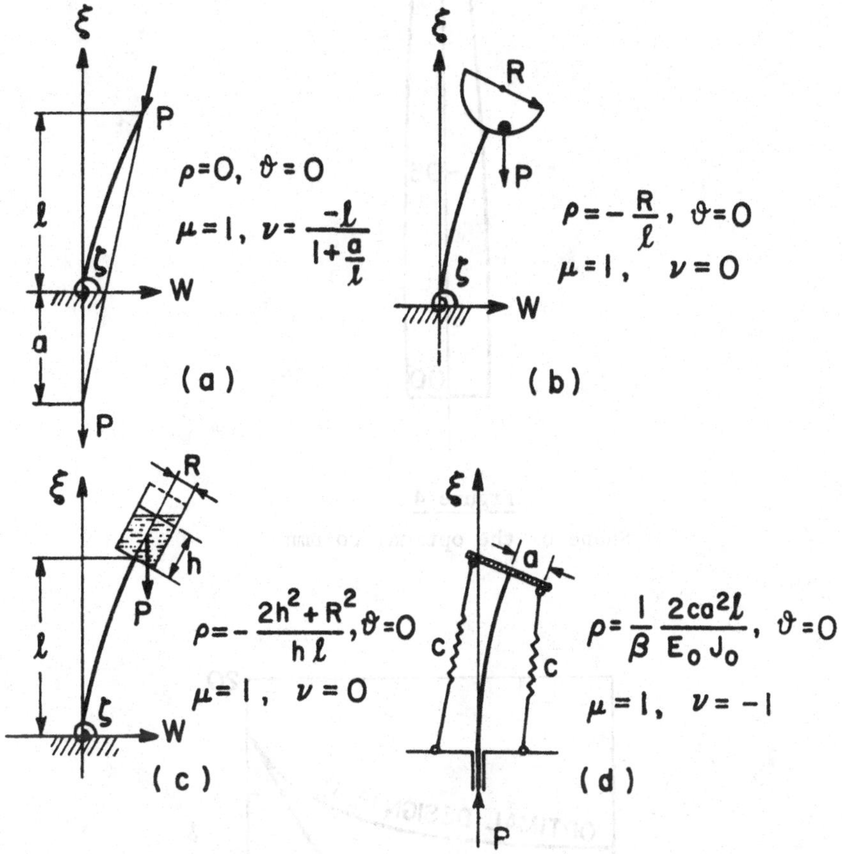

Figure 3

Various conservative loadings

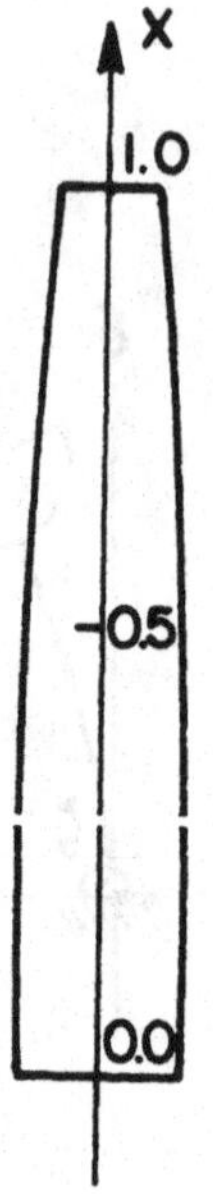

Figure 4
Shape of the optimal column

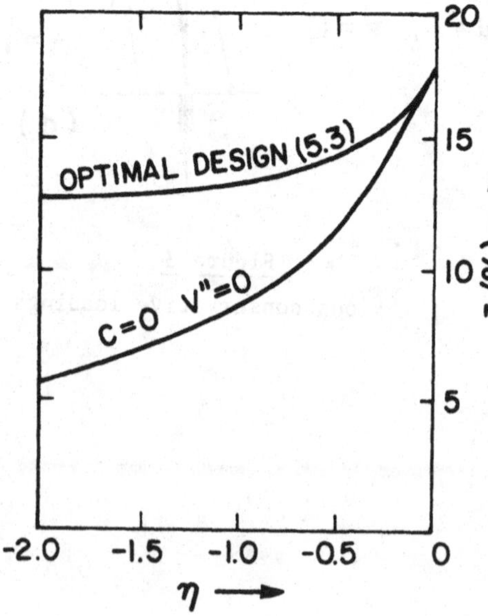

Figure 5

References, Part III, 4

[1] CLAUSEN, T., "Über die Form architektonischer Säulen", Bull. phys.-mat. de l'Académie St. Petersburg 9, 1851, pp. 368: Mélanges Math. et Astron, 1, 1849-1853, pp. 279.

[2] BLASIUS, H., "Träger kleinster Durchbiegung und Stäbe grösster Knickfestigkeit bei gegebenem Materialverbrauch", Z. Math. Phys. 62, 1914, pp. 182.

[3] NIKOLAI, E. L., "Zadatcha Lagrange'a o nayvigodneyshim otchertanii kolonn, Trudy po mekhanike", Gos. Izd. Tekh.-Teor. Lit., Moscow 1955 (first publ. 1907).

[4] TCHENTSOV, N. G., "Stoyki naymenshego vesa", Trudy TsAGI, Moscow 1936, vol. 265.

[5] TADJBAKHSH, I. and KELLER, J. B., "Strongest Columns and Isoperimetric Inequalities for Eigenvalues", J. Appl. Mech. 29, 1962, pp. 159.

[6] LAASONEN, P., "Nurjahdustuen edullisimmasta poikkipinnanvalinnasta", Tekn. Aikakauslehti 38/2, 1948, pp. 49.

[7] ZYCZKOWSKI, M., "W sprawie doboru optymalnego ksztaltu pretów osiowo sciskanych", Rozprawy Inzyn, 4/4, 1956, pp. 441.

[8] KRZYS, W., "Optymalne ksztaltowanie z uwagi na stateczność sciskanych slupow cienkosciennych o profilu zamknietym", Zeszyty Nauk. Pol. Krakowskiej, Mechanika 24, Kraków 1967.

[9] KRZYS, W., "Optimale Formen gedrückter dünnwandiger Stützen im elastisch-plastischen Bereich", Wiss. Z. TU Dresden (Internationale Stahlbautagung) 17/2, 1968, pp. 407.

[10] BECK, M., "Die Knicklast des einseitig eingespannten tangential gedrückten Stabes", Z. angew, Math. Phys. 3/3, 1952, pp. 225

[11] RZHANITSYN, A. R., "Ustoytchivost ravnovesya uprugikh system", Gos. Izd. Tekh.-Teor. Lit., Moscow 1955.

[12] KORDAS, Z., "Stability of the Elasticially Clamped Compressed Bar in the General Case of Behaviour of the Loading", Bull. Acad. Pol. Sci., Ser. Sci. Tech. 11/12, 1963, pp. 419 (English summary); Rozpr. Inzyn. 11/13, 1963, pp. 435 (Polish full text).

[13] GAJEWSKI, A. and ŻYCZKOWSKI, M., "Optimal Shaping of an Elastic Homogeneous Bar Compressed by a Polar Force", Bull. Acad. Pol. Sci., Ser. Sci. Techn. 17/10, 1969, pp. 479; 18, 1970, pp. 1 (English summaries); Rozpr. Inzyn. 17/2, 1969, pp. 299 (Polish full text).

[14] GAJEWSKI, A., "Optymalne ksztaltowanie sprezysto-plastycznego slupa przy ogolnym kouserwatywnym zachowaniu sie obciazenia", Rozpr. Inzyn. (in print).

[15] GAJEWSKI, A., "Pewne problemy optymalnego ksztaltowania preta sciskanego sila skierowana do bieguna", Mech. Teoret. Stos. 8, 1970, pp. 2.

[16] ZYCZKOWSKI, M. and GAJEWSKI, A., "Optimal Structural Design in Non-Conservative Problems of Elastic Stability", Proc. IUTAM Symp. on Instability of Continuous Systems, Herrenalb/Karlsruhe 1969 (Springer-Verlag), pp. 295-301.

[17] GAJEWSKI, A. and ZYCZKOWSKI, M., "Optimal Design of a Column Compressed by a Subtangential Force in the Elastic-Plastic Range", Arch. Mech. Stos. (in print).

[18] LAVRENTIEV, M. A. and LUSTERNIK, L. A., "Kurs variacionnogo istchislenya", Gos. Izd. Teckh.-Teor. Lit., Moscow-Leningrad 1950.

5. Investigation of Postbuckling Behaviour of Imperfect Cylindrical Shells by Means of Generalized Power Series

5.1 Introduction

The first authors dealing with the stability of shells (R. Lorenz, S. Timoshenko, R. von Mises, R. Southwell) determined upper critical pressure for geometrically perfect shells. The experiments, however, furnished as a rule much lower values of critical pressures and later two basic approaches were proposed to explain these discrepancies. The first approach, suggested by T. Kármán and H. S. Tsien [1, 2] resulted in the evaluation of lower critical pressure for perfect shells, whereas the second one, initiated by W. T. Koiter [3, 4] investigated the influence of imperfection on upper critical pressure. Extensive reviews of the literature are given in the surveys by W. A. Nash [5], Z. Nowak and M. Zyczkowski [6], B. Budiansky and J. W. Hutchinson [7], E. I. Grigolyuk and V. V. Kabanov [8], and by I. W. Hutchinson and W. T. Koiter [9].

Most recent investigations regard the upper critical pressure for imperfect shells as a suitable estimate of the limit of stability of real structures. Let us mention the papers by W. T. Koiter [10], J. W. Hutchinson [11], B. Budiansky and J. C. Amazigo [12], and many others.

Following the papers by M. Zyczkowski and K. Sobiesiak (Int. J. Non-Linear Mechanics 1973/5, 465-478; Bull. Acad. Pol. Sci., Ser. Sci. Techn. 1976/2, 53-62) we present here an analytical evaluation of upper and lower critical hydrostatic pressure for imperfect cylindrical shells of moderate length. The formulae for upper critical pressure will constitute an extension of Koiter's analysis, who determined the exponent of the first term of expansion into generalized power series; the relevant analysis for lower critical pressure has not been done before. The imperfections of the shell are assumed proportional to the buckle shape; such an assumption leads to the most conservative estimation.

5.2 Non-linear problem of stability of a circular cylindrical shell subject to hydrostatic loading

The problem under consideration was formulated by W. A. Nash [27] in 1955 and also by F. S. Isanbayeva [28] in 1955. Further investigations belong to A. S. Volmir [29], J. Kempner, K. A. V., Pandalai, S. A. Patel and J. Crouzet-Pascal [30], and to Z. Nowak [31]. Most of the authors used the energy method, obtaining the results in the form of a system of non-linear algebraic equations

$$
f_i(p,a_j,g_k,E_\ell,V_q,m,n) = 0, \qquad
\begin{aligned}
i &= 1,2,\ldots n_i, \\
j &= 1,2,\ldots n_j = n_i, \\
k &= 1,2,\ldots n_k, \\
\ell &= 1,2,\ldots n_\ell, \\
q &= 1,2,\ldots n_q,
\end{aligned}
\tag{2.1}
$$

where p denotes the loading parameter (external pressure, as a rule), a_j - the parameters of deflection (amplitudes), g_k - geometrical parameters (radius R, thickness h, length L etc.), E_ℓ - elastic constants, V_q - parameters of imperfection of the shell, and finally m and n - the numbers of circumferential and axial half-wavelengths at the length of half-perimeter of the shell. The equations of the type (2.1) will be called here "the basic formulae". To obtain an effective estimate of the influence of imperfections on the critical pressure several further operations should be carried out, namely

(1) eliminate a_j, $j = 2,3\ldots n_j$, using $n_i - 1$ equations, to obtain the relation between p and one amplitude parameter, a_1 say;

(2) solve this equation with respect to p;

(3) find the maximum (upper critical pressure) or the minimum (lower critical pressure) of the curve $p = p(a_1)$;

(4) eliminate m and n requiring the minimum (lower bound) of the critical pressure with respect to these parameters.

The last operation is obvious when evaluating the upper critical pressure, but may be subject to discussion in the case of the lower critical pressure. The question is, whether the numbers of wavelengths (particularly m, since hydrostatic pressure gives one axial half-wave anyway) may change in the course of buckling. Probably m may not change because of a certain energy barrier. Accepting this viewpoint we should substitute m corresponding to the upper critical pressure of the shell. But then we find the value of the lower critical pressure which is in a certain sense unstable with respect to m: certain other value of m gives smaller values of the lower critical pressure. To avoid this instability and to be on the safe side we may evaluate m so as to obtain the actual lower bound of the lower critical pressure. In the present paper both approaches will be used, and the two corresponding formulae for lower critical pressure will be derived and compared.

Having performed all the listed operations we obtain the sought function

$$p_{cr} = f(V_q; g_k, E_\ell), \tag{2.2}$$

determining the influence of imperfections on the critical pressure, upper or lower respectively. Usually those operations are performed numerically, using the computer. In the present paper they will be performed analytically; an appropriate approach will be used, but with relatively high accuracy for not too thick shells (such shells, however, buckle in the elastic-plastic range and this case will not be dealt with here), and not for too large imperfection parameters. Such an approach is justified since even the basic formulas are always approximate. We are not going to derive new basic formulas, but simply to apply those obtained by W. A. Nash [27], who took the imperfections of the shell into account (in contradistinction to Isanbayeva). Thus the present paper may be considered as a continuation of the paper [27], similarly as the papers [14], [13], and [32], discussing subsequently the upper critical pressure for long and for moderate-length shells, and the lower critical pressure for long shells.

Nash assumed a two-parameter deflection function

$$w(x,y) = ah\left[\sin\frac{my}{R}\sin\delta x + d(1-\cos 2\delta x)\right], \tag{2.3}$$

with the parameters $a_1 = a$ and $a_2 = d$; the constant δ refers to the number of axial half-wavelengths, $\delta = n/R$. The first term coincides with the function assumed by Mises [33] in the linear theory, whereas the second represents an additional axially-symmetrical deflection, typical of the finite-deflection theory. Minimization of the total energy with respect to the parameters a and d led to the following system of equations, describing the equilibrium of the deflected shell (quoted here after small rearrangements):

$$\frac{h^2}{24(1-\nu^2)R^2}\left[(\delta^2R^2+m^2)^2 + 32d^2R^4\delta^4\right] + \frac{h^2R^2\delta^4}{64}(2a^2+3aV+V^2)$$
$$+ d^2 + \frac{h^2m^4}{64R^2}(2a^2+3aV+V^2) - \frac{dhm^2}{8R}(3a+2V) + 2d^2h^2R^2\delta^4m^4$$
$$\times\left[\frac{1}{(\delta^2R^2+m^2)^2} + \frac{1}{(9\delta^2R^2+m^2)^2}\right](2a^2+3aV+V^2) + \frac{\delta^4R^4}{2(\delta^2R^2+m^2)^2}$$

$$\left.\begin{aligned}
&- \frac{dhR^3\delta^4m^2}{(\delta^2R^2 + m^2)^2}(3a + 2V) - \frac{pR}{4Eah}\left[m^2 - 1 - 6d^2 + \frac{\delta^2R^2}{2}(1 + 8d^2)\right] \\
&\qquad\qquad\qquad\qquad\qquad\qquad\qquad \times (2a + V) = 0, \\
&\frac{4adh^3R\delta^4}{3(1 - \nu^2)} + \frac{adh}{R} - \frac{ah^2m^2}{8R^2}(a + V) + 2adh^3R\delta^4m^4\left[\frac{1}{(\delta^2R^2 + m^2)^2}\right. \\
&\left. + \frac{1}{(9\delta^2R^2 + m^2)^2}\right](a + V)^2 - \frac{ah^2R^2\delta^4m^2}{(\delta^2R^2 + m^2)^2}(a + V) \\
&\qquad\qquad\qquad\qquad\qquad - \frac{pd}{E}(2\delta^2R^2 - 3)(a + V) = 0.
\end{aligned}\right\} \quad (2.4)$$

These equations are of the form (2.1) and will serve for our analysis as basic formulas. The parameter $V = 2a_0$ denotes the dimensionless amplitude of initial imperfections, namely this amplitude equals a_0h; distribution of the imperfections is assumed to be proportional to the deflections during buckling (most disadvantageous distribution).

5.3 Elimination of the parameter d for moderate-length shells

According to the plan presented in section 2, we have at first to eliminate the parameter $a_2 = d$ from equations (2.4). The second equation is linear with respect to d and this parameter may be evaluated exactly:

$$d = \left\{\frac{hm^2}{8R}\left[1 + \frac{8\delta^4R^4}{(\delta^2R^2 + m^2)^2}\right](a + V)\right\} : \left\{1 + \frac{4h^2R^2\delta^4}{3(1 - \nu^2)} + 2h^2R^2\delta^4m^4\left[\frac{1}{(\delta^2R^2 + m^2)^2}\right.\right.$$

$$\left.\left. + \frac{1}{(9\delta^2R^2 + m^2)^2}\right](a + V)^2 - \frac{pR}{Eah}(2\delta^2R^2 - 3)(a + V)\right\} . \quad (3.1)$$

The substitution of (3.1) into the first equation (2.4) leads, however, to extremely complicated calculations. Thus we are looking for a simpler expression for the parameter d.

We shall use the expansions with respect to the small parameter $\Phi = h/R$, or, what is identical, asymptotic expansions with respect to the large parameter

$$\lambda = \sqrt{[12(1 - \nu^2)]}\,\frac{R}{h} = 3.305\,\frac{R}{h}\,, \tag{3.2}$$

introduced in the paper [17] and called there "transversal slenderness of the shell"; the numerical value of the coefficient is given for the Poisson's ratio $\nu = 0.3$. The formula (3.1) is valid for any length of the shell; let us now confine ourselves to moderate-length shells and discuss the orders of individual terms in (3.1) with respect to the ratio h/R.

The product $\delta R = \pi R/L$ (since we put $n = \pi R/L$ corresponding to one axial semiwave) is for moderate-length shells finite, independent of Φ; similarly for the parameter of initial deflections V. The number of circumferential half-wavelengths is large; in the paper [17] devoted to the upper critical pressure of perfect shells, this number squared was found to be proportional to $\sqrt{\lambda}$ or $\Phi^{-1/2}$. In the same paper the critical pressure was found proportional to $\Phi^{5/2}$; we assume here the same orders of m^2 and p and later verify these assumptions.

A special discussion must be devoted to the order of amplitude a. In the course of buckling it starts from zero and may be of arbitrary order (not too large, because of the assumptions under which the basic formulas were derived). But the amplitudes corresponding to upper and to lower critical pressures are of a certain definite order with respect to Φ. The authors admitted in their preliminary calculations any order of a and after extremely long analysis found that (for moderate-length shells) the dimensionless amplitude corresponding to the upper critical pressure is large, of order $\Phi^{-1/3}$, and the dimensionless amplitude corresponding to the lower critical pressure is large, of order $\Phi^{-1/2}$, (the physical amplitudes ah are small of orders $\Phi^{2/3}$ and $\Phi^{1/2}$ accordingly).

To make the presentation of the analysis more compact and clear we assume here from the beginning that $a \sim \Phi^{-1/3}$ for upper critical pressure and $a \sim \Phi^{-1/2}$ for lower critical pressure, and later simply verify these assumptions. Hence the analysis of both critical pressures will be slightly different from the beginning.

Consider at first the case of upper critical pressure. Then the terms in the divident (numerator) of (3.1) are - after multiplication and necessary expansion into power series - of order 1/6, 2/6, 3/6 etc., and in the divisor (denominator) are of orders 0, 4/3 and higher. Retaining in numerator and in denominator six consecutive terms (including vanishing terms), we obtain with relatively high accuracy a simple approximation

$$d = \frac{hm^2}{8R}(a + V). \tag{3.3}$$

However, the substitution of (3.3) into the first equation (2.4) is connected with a certain complication. The basic order of this equation is 1. There are three terms of orders 2/6 and 4/6, but in their sum these orders vanish. Thus we have to rearrange these terms and to increase the accuracy in this particular case. It may easily be shown that

$$d^2 + \frac{h^2m^4}{64R^2}(2a^2 + 3aV + V^2) - \frac{dhm^2}{8R}(3a + 2V) = \left[d - \frac{hm^2}{8R}(a + V)\right] \times \left[d - \frac{hm^2}{8R}(2a + V)\right]. \tag{3.4}$$

As a matter of fact, the substitution of (3.3) into the first square bracket of (3.4) gives zero, but we have to retain the terms of orders 6/6, 7/6, 8/6 and 9/6 (four consecutive orders). To this aim we determine at first this bracket by the exact formula, making use of (3.1):

$$d - \frac{hm^2}{8R}(a + V) = \left\{\frac{hR^3\delta^4m^2}{(\delta^2R^2 + m^2)^2}(a + V) - \frac{h^2R\delta^4m^6}{4}\left[\frac{1}{(\delta^2R^2 + m^2)^2}\right.\right.$$

$$\left.\left. + \frac{1}{(9\delta^2R^2 + m^2)^2}\right](a + V)^3 - \frac{h^3R\delta^4m^2}{6(1 - \nu^2)}(a + V) + \frac{pm^2}{8Ea}(2\delta^2R^2 - 3)(a + V)^2\right\} :$$

$$\left\{1 + \frac{4h^2R^2\delta^4}{3(1 - \nu^2)} + 2h^2R^2\delta^4m^4\left[\frac{1}{(\delta^2R^2 + m^2)^2} + \frac{1}{(9\delta^2R^2 + m^2)^2}\right]\right.$$

$$\left. \times (a + V)^2 - \frac{pR}{Eah}(2\delta^2R^2 - 3)(a + V)\right\}, \tag{3.5}$$

and then retain just the terms of orders 7/6 up to 10/6. Having performed all the necessary operations, we obtain finally

$$d - \frac{hm^2}{8R}(a + V) = \frac{hR^3\delta^4}{m^2}(a + V) - \frac{1}{2}h^3R\delta^4m^2a^3 +$$

$$+ \frac{pam^2}{8E}(2\delta^2R^2 - 3) - \frac{2hR^5\delta^6a}{m^4}. \tag{3.6}$$

In the second bracket of (3.4) the accuracy of (3.3) is satisfactory.

Substituting (3.3), (3.4) and (3.6) into the first equation (2.4) and retaining the terms of orders 6/6, 7/6, 8/6 and 9/6 only, we obtain

$$\frac{h^3m^2a}{6(1 - \nu^2)R^2}(2\delta^2R^2 + m^2) - \frac{15}{8}h^3a^3R\delta^4 + \frac{2ah\delta^4R^3}{m^4}\left(1 - 2\frac{\delta^2R^2}{m^2}\right) -$$

$$- \frac{p}{E}\left[2a\left(m^2 - 1 + \frac{1}{2}\delta^2R^2\right) + m^2V\right] = 0, \tag{3.7}$$

thus completing step (1) of the analysis, adjusted to evaluation of upper critical pressure.

For lower critical pressure we proceed in a similar manner, but the discussion of orders of particular terms is slightly changed. Retaining the terms of orders 0 and 1/2 in the expression for d, (3.1), we obtain here exactly the same approximation (3.3) as before. However, the approximation for the first bracket in (3.4) is slightly changed, (more terms must be retained),

$$d - \frac{hm^2}{8R}(a + V) = \frac{ahR^3\delta^4}{m^2}\left(1 - \frac{a^2h^2m^4}{2R^2}\right) + \left[\frac{hR^3\delta^4V}{m^2}\left(1 - \frac{3a^2h^2m^4}{2R^2}\right)\right.$$

$$\left. - \frac{2ahR^5\delta^6}{m^4}\left(1 - \frac{5a^2h^2m^4}{2R^2}\right) + \frac{pam^2}{8E}(2\delta^2R^2 - 3)\right], \qquad (3.8)$$

and instead of (3.7) one obtains

$$\frac{h^2m^4}{24(1 - \nu^2)R^2} + \frac{\delta^4R^4}{2m^4} + \frac{h^2\delta^2m^2}{12(1 - \nu^2)} - \frac{\delta^6R^6}{m^6} - \frac{45ah^2R^2\delta^4V}{64} - \frac{15a^2h^2R^2\delta^4}{32}$$

$$+ \frac{a^2h^2R^4\delta^6}{m^2} + \frac{5a^3h^4\delta^4m^4V}{8} + \frac{3a^4h^4\delta^4m^4}{16} - \frac{15a^4h^4R^2\delta^6m^2}{16} - \frac{p}{aE}$$

$$\times\left[\frac{Rm^2V}{4h} + \frac{aRm^2}{2h} - \frac{aR}{2h} + \frac{a\delta^2R^3}{4h} + \frac{a^3hR\delta^2m^4}{16} - \frac{3a^3hm^4}{32R}\right] = 0. \qquad (3.9)$$

5.4 Pressure in terms of the deflection amplitude

Both equations (3.7) and (3.9) are linear with respect to the pressure p and may easily be solved. The resulting expression is of the form $p = f(a,V; m,h,R,L,E,\nu)$. For the sake of further analysis we specify the dependency of the pressure p on the amplitude α and on the imperfection parameter V and write the result in both cases as follows:

$$\frac{p}{E} = \frac{\alpha_1 a + \alpha_2 a^2V + \alpha_3 a^3 + \alpha_4 a^4V + \alpha_5 a^5}{\beta_0 V + \beta_1 a + \beta_2 a^2V + \beta_3 a^3}. \qquad (4.1)$$

The coefficients α_i and β_i may be presented in the following dimensionless form: for the analysis of upper critical pressure

$$\left.\begin{aligned} \alpha_1 &= \frac{\Phi^3 m^4}{6(1-\nu^2)} + \frac{2\phi\mu^2}{m^4} + \frac{\Phi^3\mu m^2}{3(1-\nu^2)} - \frac{4\Phi\mu^3}{m^6}, \\ \alpha_2 &= 0, \qquad \beta_0 = m^2, \\ \alpha_3 &= -\frac{15}{8}\Phi^3\mu^2, \qquad \beta_1 = 2\left(m^2 - 1 + \frac{1}{2}\mu\right), \\ \alpha_4 &= 0, \qquad \beta_2 = 0, \\ \alpha_5 &= 0, \qquad \beta_3 = 0, \end{aligned}\right\} \tag{4.2}$$

and for the analysis of lower critical pressure

$$\begin{aligned} \alpha_1 &= 32\mu^2\Phi + \frac{8\Phi^3 m^8}{3(1-\nu^2)} - 64\frac{\mu^3\Phi}{m^2} + 16\frac{\mu\Phi^3 m^6}{3(1-\nu^2)}, \\ \alpha_2 &= -45\mu^2\Phi^3 m^4, \qquad \beta_0 = 16m^6, \\ \alpha_3 &= 30\mu^2\Phi^3 m^4 + 64\mu^3\Phi^3 m^2, \qquad \beta_1 = 32m^6 - 32m^4 + 16\mu m^4, \\ \alpha_4 &= 40\mu^2\Phi^5 m^8, \qquad \beta_2 = 0, \\ \alpha_5 &= 12\mu^2\Phi^5 m^8 - 120\mu^3\Phi^5 m^6, \qquad \beta_3 = 4\mu\Phi^2 m^8 - 6\Phi^2 m^8, \end{aligned} \tag{4.3}$$

where

$$\mu = n^2 = \delta^2 R^2 = \frac{\pi^2 R^2}{L^2}. \tag{4.4}$$

5.5 Upper critical pressure

The critical pressures correspond to the extremal values of the function $p = p(a)$, namely, the upper critical pressure to a maximum, and the lower critical pressure to a minimum of this function. The analytical condition of extremum, $\partial p/\partial a = 0$, leads to the equation

$$\begin{aligned} &2\alpha_5\beta_3 a^7 + (\alpha_4\beta_3 + 3\alpha_5\beta_2)Va^6 + 2(\alpha_4\beta_2 V^2 + 2\alpha_5\beta_1)a^5 \\ &+ (-\alpha_2\beta_3 + \alpha_3\beta_2 + 3\alpha_4\beta_1 + 5\alpha_5\beta_0)Va^4 + 2(-\alpha_2\beta_3 + \alpha_3\beta_1 + 2\alpha_4\beta_0 V^2)a^3 \\ &+ (-\alpha_1\beta_2 + \alpha_2\beta_1 + 3\alpha_3\beta_0)Va^2 + 2\alpha_2\beta_0 V^2 + \alpha_1\beta_0 V = 0. \end{aligned} \tag{5.1}$$

This equation is of the seventh degree with respect to the unknown amplitude a; it determines simultaneously the amplitude of the upper critical pressure, of the lower critical pressure, and has some roots without physical meaning.

In the case of analysis of the upper critical pressure, in what follows denoted by $\bar{p}_{cr}$, several coefficients α_i and β_i vanish, (4.2), and (5.1) is reduced to

$$2\alpha_3\beta_1 a^3 + 3\alpha_3\beta_0 a^2 V + \alpha_1\beta_0 V = 0. \tag{5.2}$$

For V = 0 we obtain here the only root a = 0, corresponding just to upper critical pressure. At first let us analyse the upper critical pressure for the perfect shell V = 0 and calculate the correction for the influence of V. On substituting for α_i and β_i the relevant expressions (4.2) expanding in power series with the accuracy of four successive terms in $\Phi = h/R$ (15/6, 16/6, 17/6 and 18/6), we obtain the upper critical pressure $\bar{p}_{cr}$ for the perfect shell as a function of m, namely

$$\frac{\bar{p}_{cr}}{E} = \left[\frac{\Phi^3 m^2}{12(1-\nu^2)} + \frac{\Phi\mu^2}{m^6}\right] + \left[\frac{(3\mu+2)\Phi^3}{24(1-\nu^2)} + \frac{(2-5\mu)\Phi\mu^2}{2m^8}\right]. \tag{5.3}$$

The extremum (minimum) condition in m^2, i.e. $\partial\bar{p}_{cr}/\partial(m^2) = 0$, leads to the equation

$$\Phi^2 m^{10} - 36(1-\nu^2)\mu^2 m^2 - 24(1-\nu^2)\mu^2(2-5\mu) = 0, \tag{5.4}$$

the solution of which yields (for p_{min})

$$m^2 = U = \sqrt{(\mu\lambda\sqrt{3})} - \left((5/6)\mu - \frac{1}{3}\right) + \ldots = \sqrt{6}\,\sqrt[4]{(1-\nu^2)}\,\sqrt{\mu/\Phi} - \left((5/6)\mu - \frac{1}{3}\right) + 0\cdot(\sqrt{\Phi}), \tag{5.5}$$

where the symbol $0(\sqrt{\Phi})$ indicates the order of the first eliminated term.

On substituting (5.5) into (5.3) we obtain an explicit formula for the upper critical load for a perfect shell of medium length with an accuracy of four successive orders (counted every 1/6)

$$\frac{\bar{p}_{cr}}{E} = \frac{\sqrt{6}}{9(1-\nu^2)^{3/4}}\,\mu^{1/2}\,\Phi^{5/2} + \frac{\mu + 2}{18(1-\nu^2)}\,\Phi^3 . \tag{5.6}$$

To take into account the influence of the parameter V, let us again consider equation (5.2). We are interested here in the upper critical pressure, and we seek for the extremum value of p, for V = 0 and a = 0. To obtain an analytic solution for imperfect shell let us consider V to constitute a new small parameter and represent the solution of equation (5.2) in the form of a generalized power series

$$a = \bar{a} = \sum_{j=0}^{\infty} a_j\, V^{\mu+\nu j} , \tag{5.7}$$

where the exponents μ, $\mu + \nu$, $\mu + 2\nu$,... are unknown, but $\mu \neq 0$. The values $\mu = 1/3$ and $\nu = 2/3$ satisfy the conditions (5.2). Another combination, $\mu = 0$ and $\nu = 1$, corresponds to the lower critical pressure and will be analysed later.

On substituting into (5.2) the series

$$a = \bar{a} = a_0 V^{1/3} + a_1 V + a_2 V^{5/3} + \ldots , \tag{5.8}$$

and setting equal to zero the coefficients of V, $V^{5/3}$, $V^{7/3}$, ..., we obtain a system of equations for a_0, a_1, a_2,...

$$\alpha_1\beta_0 + 2\alpha_3\beta_1 a_0^3 = 0, \qquad \beta_0 + 2\beta_1 a_1 = 0, \tag{5.9}$$

$$\beta_0 a_1 + \beta_1(a_0 a_2 + a_1^2) = 0, \ldots$$

Hence, confining ourselves to three coefficients and substituting them into (5.8) we obtain the amplitude (as expressed in terms of α_i, and β_i) corresponding to $\bar{p}_{cr}$ for an imperfect shell of medium length, in the form

$$a = \bar{a} = \sqrt[3]{\frac{-\alpha_1\beta_0}{2\alpha_3\beta_1}}\, V^{1/3} - \frac{\beta_0}{2\beta_1} V + \frac{\beta_0^2}{4\beta_1^2} \sqrt[3]{\frac{2\alpha_3\beta_1}{-\alpha_1\beta_0}}\, V^{5/3} - \dots \tag{5.10}$$

The series (5.8) which has already been determined, is now substituted into the expression (4.1) for the pressure. After some transformations, we find

$$\frac{\bar{p}_{cr}}{E} = \frac{\alpha_1}{\beta_1}\left(1 - \frac{3}{2}\sqrt[3]{\frac{-2\alpha_3\beta_0^2}{\alpha_1\beta_1^2}}\, V^{2/3} + 3\sqrt[3]{\frac{\alpha_3^2\beta_0^4}{2\alpha_1^2\beta_1^4}}\, V^{4/3} - \dots\right) \tag{5.11}$$

This is a general relation for $\bar{p}_{cr}$ for imperfect (real) shells of moderate length under hydrostatic pressure. For V = 0 it becomes the Mises formula for a perfect shell of moderate length in the function of m, obtained on the grounds of the linear theory.

Let us now consider the parameters $\bar{p}_{cr}$ for an imperfect shell of moderate length in an explicit form. We shall examine the amplitude a for p_{cr}. If we substitute into (5.10) equation (4.2) and perform the required expansions (with the same accuracy as in Sec. 3) in power series of $\Phi = h/r$, we shall find for the amplitude a, corresponding to p_{cr} in function of m

$$a = \bar{a} = \sqrt[3]{\frac{2}{15\Phi^3\mu^2}\left(\frac{\Phi^3 m^4}{6(1-\nu^2)} + \frac{2\Phi\mu^2}{m^4}\right)} \times$$

$$\times\left[1 + \frac{1}{3}\left(\frac{2\Phi^2\mu m^8 - 24(1-\nu^2)\mu^3}{\Phi^2 m^{10} + 12(1-\nu^1)\mu^2 m^2} + \frac{2-\mu}{2m^2}\right)\right] V^{1/3} - \frac{1}{4} V +$$

$$+ \sqrt[3]{\frac{45(1-\nu^2)\Phi^2\mu^2 m^4}{4096\Phi^2 m^8 + 49152(1-\nu^2)\mu^2}}\, V^{5/3} - \dots \tag{5.12}$$

The relation (5.12) has been established by retaining the terms of order -2/6, -1/6, 0, 1/6 and 2/6. The fifth term of order 2/6 is, for $V^{5/3}$, beyond the accuracy required (four successive terms with regard to $\Phi = h/R$). It has been retained in order to emphasize the general structure of the generalized power series for $a = \bar{a}$. It is found that for the effective determination of the amplitude (of an imperfect shell of moderate length) it suffices to substitute for $m^2 = U$ into (5.12) the expression (5.5) which is valid for the perfect shell. This substitution leads to the equation

$$a = \bar{a} = \left[\frac{2\sqrt[3]{3/5}\,\mu\Phi}{3\sqrt[6]{(1-\nu^2)}} + \frac{\sqrt{6}\,\sqrt[3]{3/5}}{81\sqrt[12]{(1-\nu^2)^5}}(4-\mu)\sqrt[6]{\frac{\Phi}{\mu^5}}\right]V^{1/3} -$$

$$- \frac{1}{4}\Phi^0 V + \frac{3}{32}\sqrt[3]{\frac{5/\mu\Phi}{3}}\sqrt[6]{(1-\nu^2)}\,V^{5/3} - \ldots \qquad (5.13)$$

the term of order 2/6 (for $V^{5/3}$) being given for the sake of information.

Proceeding we analyze now the influence of V on $\bar{p}_{cr}$. In order to establish an explicit formula we substitute (4.2) into (5.11) and expand into series of $\Phi = h/R$ leaving four successive terms, and additionally the fifth term, which is done for for the same reasons as in the analysis of the amplitude a. Thus by retaining the terms of orders 15/6 to 18/6 and in the last term of the series of order 19/6 we obtain

$$\frac{\bar{p}_{cr}}{E} = \left[\frac{\Phi^3 m^2}{12(1-\nu^2)} + \frac{\Phi\mu^2}{m^6}\right] + \left[\frac{(3\mu+2)\Phi^3}{24(1-\nu^2)} + \frac{(2-5\mu)\Phi\mu^2}{2m^8}\right] -$$

$$- \frac{3}{2}\sqrt[3]{\frac{15\Phi^3\mu^2}{128m^6}\left(\frac{\Phi^3 m^4}{6(1-\nu^2)} + \frac{2\Phi\mu^2}{m^4}\right)^2}\,V^{2/3} +$$

$$+ 3\sqrt[3]{\left(\frac{15\Phi^3\mu^2}{128m^3}\right)^2\left(\frac{\Phi^3 m^4}{6(1-\nu^2)} + \frac{2\Phi\mu^2}{m^4}\right)}\,V^{4/3} - \ldots \qquad (5.14)$$

To eliminate m from (5.14) (to establish $\bar{p}_{cr\ min}$ in function of m) we write the extremum (minimum) condition.

This condition yields

$$4\sqrt[3]{225\mu\Phi}[\Phi^2m^8 + 12(1-\nu^2)\mu^2]^4\ [\Phi^2m^{10} - 36(1-\nu^2)\mu^2m^2 -$$

$$- 24(1-\nu^2)\mu^2(2-5\mu)] - 15\Phi\mu\sqrt[3]{3(1-\nu^2)m^{10}}[\Phi^4m^{16} -$$

$$- 72(1-\nu^2)\Phi^2\mu^2m^8 - 1008(1-\nu^2)^2\mu^4]V^{2/3} + \ldots = 0. \qquad (5.15)$$

The solution (5.15) can be given in the form of a double series of powers of the parameters Φ and V

$$m^2 = U = \sum_{j=0}^{\infty}\sum_{n=0}^{\infty} C_{jn}\,\Phi^{(j-1)/2}\,V^{2n/3}, \qquad (5.16)$$

that is

$$m^2 = U = \sqrt[4]{(1-\nu^2)}\sqrt{\frac{6\mu}{\Phi}} - \left(\frac{5}{6}\mu - \frac{1}{3}\right) + 0\cdot(\sqrt{\Phi}) -$$

$$- \frac{5\sqrt{6}}{24}\sqrt[3]{\frac{9}{25}}\sqrt[12]{(1-\nu^2)^5}\sqrt[6]{\frac{\mu^5}{\Phi}}\cdot V^{2/3} + \ldots \qquad (5.17)$$

The substitution of (5.17) into (5.14) yields the required minimum value of $\bar{p}_{cr}$ with respect to m

$$\frac{\bar{p}_{cr}}{E} = \frac{\sqrt{6\mu}}{9\sqrt[4]{(1-\nu^2)^3}}\Phi^{15/6} + \frac{\mu+2}{18(1-\nu^2)}\Phi^{18/6} + \ldots - \frac{1}{8}\sqrt[3]{10\sqrt{6}}(1-\nu^2)^{-7/12}\ \times$$

$$\times\ \mu^{5/6}\,\Phi^{17/6}\,V^{2/3} + \frac{3}{32}\sqrt[3]{\frac{50\sqrt{6}}{3(1-\nu^2)}}(1-\nu^2)^{-1/12}\,\mu^{7/6}\,\Phi^{19/6}\,V^{4/3} - \ldots \qquad (5.18)$$

Accordingly, we can now verify our assumptions for the orders of the magnitude of m^2, $\bar{p}_{cr}$ and $a = \bar{a}$. In Sec. 3 the fundamental orders of the magnitude were assumed thus: $m^2 \sim \Phi^{-1/2}$, $\bar{p}_{cr} \sim \Phi^{5/2}$ and $a \sim \Phi^{1/3}$. These assumptions are verified by the series (5.17), (5.18) and (5.13). From (5.18) it follows that the imperfection parameter V leads to a relatively sharp decrease in $\bar{p}_{cr}$ the main correcting term being in direct proportion to $V^{2/3}$, as it was shown by W. T. Koiter [3, 4].

5.6 Lower critical pressure

Return once more to the condition of extremum of pressure p, (5.1). Similarly as in the preceding case we start with the analysis of a perfect shell, V = 0, and later evaluate the corrections due to imperfections.

For V = 0 equation (5.1) takes the form

$$a^3[\alpha_5\beta_3 a^4 + 2\alpha_5\beta_1 a^2 + (-\alpha_1\beta_3 + \alpha_3\beta_1)] = 0. \tag{6.1}$$

The root a = 0 corresponds here to the upper critical pressure; the lower critical pressure will be reached for one of the roots of the biquadratic equation in the square bracket. Note that the terms $2\alpha_5\beta_1 a^2$ and $\alpha_3\beta_1$ are - according to the estimation given in section 3 - of order 1/2, whereas the remaining terms are of order 1. Thus in the basic approximation, for very thin shells, the amplitude corresponding to the lower critical pressure is determined by the formula

$$a^2 = -\frac{\alpha_3}{2\alpha_5} \tag{6.2}$$

[α_3, according to (4.3), is here negative]. This accuracy is already sufficient to obtain two accurate terms in the formula for the critical pressure. More exactly, a^2 may be evaluated as the appropriate root of (6.1), namely

$$a^2 = \frac{\alpha_1\beta_3 - \alpha_3\beta_1}{\alpha_5\beta_1 + \sqrt{(\alpha_5^2\beta_1^2 + \alpha_1\alpha_5\beta_3^2 - \alpha_3\alpha_5\beta_1\beta_3)}} . \tag{6.3}$$

Making use of the difference in orders of individual coefficients α_i and β_i we may also perform the expansion of (6.3):

$$a^2 = -\frac{\alpha_3}{2\alpha_5}\left[1 - \left(\frac{\alpha_1}{\alpha_3} - \frac{\alpha_3}{4\alpha_5}\right)\frac{\beta_3}{\beta_1} + \dots\right], \tag{6.4}$$

or, with (4.3) taken into account,

$$a^2 = \frac{5}{4}\Phi^{-2}m^{-4}\left\{1 + \left[\frac{507}{64}\frac{\mu}{m^2} - \frac{53}{640}\frac{1}{m^2} + \frac{\Phi^2 m^6}{90(1-\nu^2)\mu} - \frac{\Phi^2 m^6}{60(1-\nu^2)\mu^2}\right] + ..\right\} \quad (6.5)$$

It may be proved that this amplitude corresponds to the minimum of the curve $p = p(a)$. Substituting (6.3) or (6.5) into (4.1) with $V = 0$ we determine the lower critical pressure $\bar{\bar{p}}_{cr}$ for a perfect shell as a function of the number of circumferential half-wave lengths m. Some simple transformations yield

$$\frac{\bar{\bar{p}}_{cr}}{E} = \frac{4\alpha_1\alpha_5 - \alpha_3^2}{2\beta_1\alpha_5 - \alpha_3\beta_3 + 2\sqrt{[\alpha_5(\alpha_5\beta_1^2 + \alpha_1\beta_3^2 - \alpha_3\beta_1\beta_3)]}} \quad (6.6)$$

and after expansions

$$\frac{\bar{\bar{p}}_{cr}}{E} = \frac{4\alpha_1\alpha_5 - \alpha_3^2}{4\alpha_5\beta_1}\left(1 + \frac{\alpha_3\beta_3}{2\alpha_5\beta_1} + \ldots\right), \quad (6.7)$$

$$\frac{\bar{\bar{p}}_{cr}}{E} = \left(\frac{53}{128}\mu^2\Phi m^{-6} + \frac{1}{12}\frac{\Phi^3 m^2}{1-\nu^2}\right) + \left(-\frac{23065}{4096}\mu^3\Phi m^{-8} + \frac{4187}{8192}\mu^2\Phi m^{-8} + \frac{43}{384}\frac{\mu\Phi^3}{1-\nu^2} + \frac{79}{768}\frac{\Phi^3}{1-\nu^2}\right) + \ldots \quad (6.8)$$

the brackets in (6.8) collect terms of the same order.

There arises now the problem of the appropriate choice of the number of circumferential half-lengths m. This problem was discussed in section 2. We derive and compare here final formulae for both approaches: (1) m constant during buckling, thus corresponding to the upper critical pressure $\bar{p}_{cr}$; (2) m furnishing a minimal value of the lower critical pressure.

The calculations will be carried out for imperfect shells, thus we return at first to the equation (5.1). To obtain its analytical solution we consider the imperfection parameter V as a new small parameter and present the root in the form

$$a = \sum_{j=0}^{\infty} a_j(\Phi,\mu,\nu;m)V^j , \tag{6.9}$$

where a_0 is determined by (6.3), (6.4), or (6.5) substituting (6.9) into (5.1) and equating the coefficients of V^1 to zero we find

$$a_1 = \frac{-\alpha_4\beta_3 a_0^6 + \alpha_2\beta_3 a_0^4 - 3\alpha_4\beta_1 a_0^4 - 5\alpha_5\beta_0 a^4 - \alpha_2\beta_1 a_0^2 - 3\alpha_3\beta_0 a_0^2 - \alpha_1\beta_0}{14\alpha_5\beta_3 a_0^6 + 10\alpha_5\beta_1 a^4{}_0 - 6\alpha_1\beta_3 a_0^2 + 6\alpha_3\beta_1 a_0^2} . \tag{6.10}$$

Retaining for a_0 only the basic approximation, (6.2), and omitting the terms of order higher than (-1/2)(with respect to Φ, according our preliminary estimations) both in the numerator and in the denominator we obtain

$$a_1 = \frac{3\alpha_4}{2\alpha_5} - \frac{\beta_0}{2\beta_1} - \frac{\alpha_2}{\alpha_3} + \frac{2\alpha_1\alpha_5\beta_0}{\alpha_3^2\beta_1} , \tag{6.11}$$

or with (4.3) substituted and consequently neglecting higher-order terms,

$$a_1 = \frac{1103}{300} + \frac{8}{225}\frac{\Phi^2 m^8}{(1-\nu^2)\mu^2} . \tag{6.12}$$

Thus the expansion (6.9) has the form of an ordinary power series with respect to V, whereas the similar expansion determining the amplitude of the upper critical pressure is given by a generalized power series, Sec. 5. Combining (6.5) and (6.12) we determine the deflection amplitude corresponding to the lower critical pressure by the formula

$$a = \frac{\sqrt{5}}{2}\Phi^{-1}m^{-2}\left\{1 + \left[\frac{507}{128}\frac{\mu}{m^2} - \frac{53}{1280}\frac{1}{m^2} + \frac{\Phi^2 m^6}{180(1-\nu^2)\mu} - \frac{\Phi^2 m^6}{120(1-\nu^2)\mu^2}\right] + \ldots\right\}$$
$$+ \left[\frac{1103}{300} + \frac{8}{225}\frac{\Phi^2 m^8}{(1-\nu^2)\mu^2}\right] V + \ldots \tag{6.13}$$

It turns out that the imperfections result in an increase of this amplitude; the influence of the parameter Φ goes in the same direction, since μ for moderate-length shells is sufficiently large and the correction is positive.

To obtain the formula for the lower critical pressure we substitute (6.13) into (4.1) and perform the expansions with respect to Φ and to V. The substitution yields at first (the orders of α_i, and β_i being estimated from (4.3)):

$$\frac{\bar{\bar{p}}_{cr}}{E} = \frac{4\alpha_1\alpha_5 - \alpha_3^2}{4\alpha_5\beta_1} + \frac{4\alpha_1\alpha_3\alpha_5\beta_3 - \alpha_3^3\beta_3}{8\alpha_5^2\beta_1^2}$$

$$+ \left(\frac{\alpha_2}{\beta_1} - \frac{\alpha_3\alpha_4}{2\alpha_5\beta_1} + \frac{2\alpha_1\alpha_5\beta_0}{\alpha_3\beta_1^2} - \frac{\alpha_3\beta_0}{2\beta_1^2}\right) V \sqrt{\left(-\frac{\alpha_3}{2\alpha_5} + \ldots\right)}, \tag{6.14}$$

and with (4.3) taken into account

$$\frac{\bar{\bar{p}}_{cr}}{E} = \left(\frac{53}{128}\mu^2\Phi m^{-6} + \frac{1}{12}\frac{\Phi^3 m^2}{1 - \nu^2}\right) + \left(-\frac{23065}{4096}\mu^2\Phi m^{-8} + \frac{4187}{8192}\mu^2\Phi m^{-8}\right.$$

$$\left. + \frac{43}{384}\frac{\mu\Phi^3}{1 - \nu^2} + \frac{79}{768}\frac{\Phi^3}{1 - \nu^2}\right) - \left(\frac{3\sqrt{5}}{640}\mu^2\Phi^2 m^{-4} + \frac{\sqrt{5}}{60}\frac{\Phi^4 m^4}{1 - \nu^2}\right) V + \ldots \tag{6.15}$$

Assume now that the number of circumferential half-wavelengths m is constant during buckling. If we assume that for the lower critical pressure m^2 is determined by (5.17) as for the upper critical pressure, and substitute (5.17) into (6.15), then this pressure equals

$$\frac{\bar{\bar{p}}_{cr}}{E} = \frac{437\sqrt{6}}{4608}\frac{\mu^{1/2}\Phi^{5/2}}{(1 - \nu^2)^{3/4}} - \frac{(25106\mu - 39323)\Phi^3}{294912(1 - \nu^2)} - \frac{129\sqrt{5}}{1280}\frac{\mu\Phi^3}{\sqrt{(1 - \nu^2)}} V + \ldots \tag{6.16}$$

and the corresponding dimensionless amplitude (6.13) is

$$a = \frac{\sqrt{30}}{12^4\sqrt{(1 - \nu^2)}\sqrt{(\mu\Phi)}} + \frac{(19178\mu - 2591)\sqrt{5}}{46080\mu\sqrt{(1 - \nu^2)}} + \frac{1487}{300} V + \ldots \tag{6.17}$$

The second, more conservative alternative consists in the evaluation of minimal $\bar{\bar{p}}_{cr}$ with respect to m. It is seen that (6.15) has a minimum; the condition $\partial\bar{\bar{p}}_{cr}/\partial m = 0$ leads to the equation

$$\left[-\frac{159}{64}\mu^2\Phi m^{-7} + \frac{\Phi^3 m}{6(1-\nu^2)}\right] + \left(\frac{23065}{512}\mu^3\Phi m^{-9} - \frac{4187}{1024}\nu^2\Phi m^{-9}\right) + \left(\frac{3\sqrt{5}}{160}\mu^2\Phi^2 m^{-5} - \frac{\sqrt{5}}{15}\frac{\Phi^4 m^3}{1-\nu^2}\right)V + \ldots = 0. \tag{6.18}$$

Its solution may be presented in the form of a double power series of the form

$$m^2 = U = \sum_{j=0}^{\infty}\sum_{n=0}^{\infty} c_{jn}\,\Phi^{(j-1)/2}\,V^n, \tag{6.19}$$

namely

$$m^2 = U = \sqrt[4]{\left[\frac{477}{32}(1-\nu^2)\right]}\sqrt{\frac{\mu}{\Phi}} - \left(\frac{23065}{5088}\mu - \frac{79}{192}\right) + \frac{39}{20}\sqrt{\left[\frac{10}{53}(1-\nu^2)\right]}\mu V + . \tag{6.20}$$

Substitution of (6.20) into (6.15) yields the sought critical pressure, minimal with respect to m:

$$\frac{\bar{\bar{p}}_{cr}}{E} = \frac{53}{32}\left[\frac{32}{477(1-\nu^2)}\right]^{3/4}\mu^{1/2}\Phi^{5/2} - \frac{(8114\mu - 4187)\Phi^3}{30528(1-\nu^2)} - \frac{27\mu\Phi^3}{8\sqrt{530}(1-\nu^2)}V + .. \tag{6.21}$$

The corresponding dimensionless amplitude (6.13) with substituted (6.20) equals

$$a = \sqrt[4]{\left[\frac{50}{477(1-\nu^2)}\right]}\frac{1}{\sqrt{(\mu\Phi)}} + \frac{(218198\mu - 14681)}{12720\mu}\sqrt{\left[\frac{10}{477(1-\nu^2)}\right]} + \frac{631}{150}V + . \tag{6.22}$$

Now we may check our assumptions as regards the orders of m , $\bar{\bar{p}}_{cr}$, and a. In section 3 we assumed the basic orders as follows: $m^2 \sim \Phi^{-1/2}$, $\bar{\bar{p}}_{cr} \sim \Phi^{5/2}$, $a \sim \Phi^{-1/2}$. The series (6.16), and (6.17), or the series (6.20), (6.24), and (6.22) verify clearly these assumptions.

It is interesting to compare the series (6.16) and (6.21) determining $\bar{\bar{p}}_{cr}$ based on various concepts as regards the number of circumferential half-wave lengths m. The basic term of (6.21) is about 6 per cent lower; for larger value of the ratio $h/R=\Phi$ the difference is even higher, but, on the other hand, the imperfection parameter V results in a decrease of the numerical differences between both formulae.

5.7 Summary of results written in physical quantities

The series given above describe the parameters of upper and lower critical pressure in dimensionless form. Going back to the physical quantities, more convenient in engineering applications we present them finally in the form (for $\nu = 0.3$): upper critical pressure

$$m^2 = 7.51214 \frac{R}{L}\sqrt{\frac{R}{h}} - \left(8.22434 \frac{R^2}{L^2} - 0.3333\right) - $$

$$- 3.71862 \frac{R^2}{Lh} \sqrt[6]{\frac{h}{R}} \sqrt[3]{\frac{f_0^2}{L^2}} + \dots$$

$$\frac{\bar{p}_{cr}}{E} = 0.91726 \frac{h^2}{LR}\sqrt{\frac{h}{R}} + 0.1221\left(1 + \frac{\pi^2 R^2}{2L^2}\right)\left(\frac{h}{R}\right)^3 - \tag{7.1}$$

$$- 4.13884 \frac{h^2}{LR} \sqrt[3]{\left(\frac{h}{L}\right)^2} \sqrt[6]{\frac{h}{R}} \sqrt[3]{\frac{f_0^2}{h^2}} + \dots$$

$$f = \left[0.2668 \sqrt[3]{\frac{L^2}{Rh}} + 0.0105 \frac{4L^2 - \pi^2 R^2}{LR} \sqrt[3]{\frac{L^2}{R^2}} \sqrt[6]{\frac{h}{R}}\right] \sqrt[3]{2h^2 f_0} - \frac{1}{2} f_0 + \dots,$$

and more conservative estimation of lower critical pressure

$$m^2 = 6.029\,\frac{R}{L}\sqrt{\left(\frac{R}{h}\right)} - \left(44.74\,\frac{R^2}{L^2} - 0.412\right) + 15.95\,\frac{R^2}{L^2}\,\frac{f_0}{h} + \ldots$$

$$\frac{\bar{\bar{p}}_{cr}}{E} = 0.7361\,\frac{R}{L}\left(\frac{h}{R}\right)^{5/2} - \left(2.883\,\frac{R^2}{L^2} - 0.151\right)\left(\frac{h}{R}\right)^3 - 3.034\,\frac{h^2 f_0}{L^2 R} + \ldots \quad (7.2)$$

$$f = 0.1854L\sqrt{\left(\frac{h}{R}\right)} + \left(2.604 - 0.0178\,\frac{L^2}{R^2}\right)h + 8.414\,f_0 + \ldots,$$

where $f = ah$ and $f_0 = a_0 h$ are the physical amplitudes of the deflections.

It may be seen that the differences between the upper and the lower critical pressure decrease with the increase of imperfections, since $\bar{p}_{cr}$ decreases more rapidly (with $V^{2/3}$) whereas $\bar{\bar{p}}_{cr}$ much slower (with V^1).

References, Part III, 5

[1] KARMAN, TH. and TSIEN, H. S., "The buckling of thin spherical shells by external pressure", J. Aero Sci. 7, 1939, pp. 43-50.

[2] KARMAN, TH. and TSIEN, H. S., "The buckling of thin cylindrical shells under axial compression", J. Aero. Sci. 8, 1941, pp. 303.

[3] KOITER, W. T., "On the stability of elastic equilibrium (in Dutch). Thesis, Amsterdam (1945) (English translation: NASA TT F-10, 1967, pp. 833).

[4] KOITER, W. T., "Elastic stability and postbuckling behavior", Proc. Symp. Non-linear Problems, Univ. of Wisconsin Press, Madison, 1963, pp. 257-275.

[5] NASH, W. A., "Recent advances in the buckling of thin shells", Appl. Mech. Rev. 13, 1960, pp. 161-164.

[6] NOWAK, Z. and ŻYCZKOWSKI, M., "Review of recent papers on the stability of thin shells" (in Polish), Mechanika Teor. Stos. 1,2, 1963, pp. 31-66.

[7] BUDIANSKY B. and HUTCHINSON, J. W., "A survey of some buckling problems", AIAA J. 4, 1966, pp. 1505-1510.

[8] GRIGOLYUK, E. I. and KABANOV, V. V., "Stability of circular cylindrical shells" (in Russian), Itogi Nauki, Seria Mekhanika, Moscow, 1969.

[9] HUTCHINSON, J. W. and KOITER, W. T., "Postbuckling theory", Appl. Mech. Rev. 23, 1970, pp. 1353-1366.

[10] KOITER, W. T., "The effect of axisymmetric imperfections on the buckling of cylindrical shells under axial compression", Koninkl. Ned. Akad. Wetenschap., Proc. Ser. B66, 1963, pp. 265-279; Lockheed Miss. Space Co. TR 6-90-63-86, August 1963.

[11] HUTCHINSON, J. W., "Axial buckling of pressurized imperfect cylindrical shells", AIAA J. 3, 1965, pp. 1461-1466.

[12] BUDIANSKY, B. and AMAZIGO, J. C., "Initial postbuckling behavior of cylindrical shells under external pressure", J. Math. Phys. 47, 1968, pp. 223-235.

[13] SOBIESIAK, K., "Analytical treatment of the influence of imperfections on upper critical pressure for moderate-length cylindrical shells", (in preparation).

[14] ŻYCZKOWSKI, M. and SOBIESIAK, K., "Analytical treatment of the influence of imperfections on upper critical pressure for long cylindrical shells", Arch. Mech. Stosowanej 23, 1971, pp. 875-884.

[15] ROORDA, J., "The buckling behavior of Imperfect Structural Systems", J. Mech. Phys. Solids 13, 1965, pp. 267-280.

[16] ROORDA, J., "On the buckling of symmetric structural systems with first and second order imperfections", Int. J. Solids Structures 4, 1968, pp. 1137-1148.

[17] ŻYCZKOWSKI, M., "Analysis of stability of radially compressed cylindrical shells by means of generalized power series", Conference Theory of Plates and Shells, SAV, 505-514, Bratislava (1966) (English summary); Rozpr. Inz. 14, 1966, pp. 157-174 (Polish full text).

[18] BUCKO, S., "Analysis of stability of axially compressed cylindrical shells by the method of generalized power series", Bull. Acad. Pol. Sci., Ser. Sci. Techn. 15, 1967, pp. 303-312, (English Summary); Archiwum Budowy Maszyn 13, 1966, pp. 307-327 (Polish full text).

[19] ŻYCZKOWSKI, M. and BUCKO, S., "A method of stability analysis of cylindrical shells under biaxial compression", AIAA J. 9, 1971, pp. 2259-2263.

[20] SEWELL, M. J., "The static perturbation technique in buckling problems", J. Mech. Phys. Solids 13, 1965, pp. 247-265.

[21] SEWELL, M. J., "A method of postbuckling analysis", J. Mech. Phys. Solids 17, 1969, pp. 219-233.

[22] THOMPSON, J. M. T., "A new approach to elastic branching analysis", J. Mech. Phys. Solids 18, 1970, pp. 29-42.

[23] DYM, C. L. and HOFF, N. J., "Perturbation solutions for the buckling problems of axially compressed thin cylindrical shells of finite or infinite length", Trans. ASME E35, 1968, pp. 754-762.

[24] KAYUK, YA. F., "On the convergence of the small parameter expansions in the geometrically nonlinear problems", (in Russian), Prikladnaya Mekhanika 9, 3, 1973, pp. 83-89.

[25] POGORELOV, A. V., "Geometrical theory of stability of shells" (in Russian), izd. Nauka, Fiz.-Mat. Lit., Moscow 1966.

[26] HOFF, N. J., MADSEN, W. A. and MAYERS, J., "Postbuckling equilibrium of axially compressed circular cylindrical shells", AIAA J. 4, 1966, pp. 126-133.

[27] NASH, B. A., "Effect of large deflections and initial imperfections on the buckling of cylindrical shells subject to hydrostatic pressure", J. Aero. Sci. 22, 1955, pp. 264-269.

[28] ISANBAYEVA, F. S., "Evaluation of lower critical pressure of a cylindrical shell under hydrostatic pressure" (in Russian), Izv. kazan. Fil. Akad. Nauk SSSR 7, 1955, pp. 51-58.

[29] VOLMIR, A. S., "On the influence of imperfections on the stability of cylindrical shells subject to hydrostatic pressure" (in Russian), Dokl. Akad. Nauk SSSR 113, 1957, pp. 291-293.

[30] KEMPNER, J., PANDALAI, K. A. V., PATEL, S. A. and CROUZET-PASCAL, J., "Postbuckling behavior of circular cylindrical shells under hydrostatic pressure", J. Aero. Sci. 24, 1959, pp. 253--264.

[31] NOWAK, Z., "Non-linear problem of stability of a closed orthotropic cylindrical shell with clamped edges", Bull. Acad. Polon. Sci., Ser. Sci. Techn. 12, 1964, pp. 165-175 (English summary); Arch. Budowy Maszyn 11, 1964, pp. 619-636 (Polish full text).

[32] SOBIESIAK, K., "Analytical evaluation of lower critical hydrostatic pressure for imperfect long cylindrical shells" (in preparation).

[33] MISES, R., "Der kritische Aussendruck für allseits belastete cylindrische Rohre", Festschr. zum 70. Geburtstag von Prof. Dr. A. Stodola, Zürich, 1929.

[34] ZYCZKOWSKI, M., "Operations on generalized power series", Z. angew. Math. Mechanik 45, 1965, pp. 235-244.

6. Optimal Design of Shells with Respect to Their Elastic Stability

6.1 Introductory remarks

Thin-walled shells are often subject to the loss of stability and then the corresponding constraint should be taken into account in their structural optimization.

The stability of shells may be improved by reinforcing elements, e.g. ribs, pegs, cables, etc. Many papers deal with the optimal design of such reinforcements; let us mention here the papers by A. B. Burns [4], A. B. Burns and J. Skogh [5], O. I. Terebushko [19], J. Singer and M. Baruch [17], L. Spunt [18], M. Zyczkowski [25] (optimal design of point-reinforcement) and V. I. Korolev [11] (optimal design of reinforcement of plastics). Reinforced shells may often be regarded as shells the thickness of which is given by generalized functions (distributions). However, for some technological reasons, the reinforcing elements may be found inconvenient, and the present paper will in principle be confined to the shells with the thickness described by smooth functions.

Optimization of shell structures presents the problems which are fairly rich in design variables. According to the classification given in [28], the typical design variables may be related to the thickness of the shell, the shape of the mid-surface, the shape of the boundary and to the mode of support of the shell. A survey of shell optimization is given by V. V. Vasilev [21], but the section connected with the stability constraints is there very scarce. On the other hand, R. H. Plaut [16] in his review of optimal design with stability constraints pays practically no attention to shell optimization. Some related problems are mentioned in the survey by M. Zyczkowski [27], namely the optimization of thin-walled bars in which the stability constraints are formulated on the basis of the shell theory (M. Feigen [6], W. Krzyś [12]). Parametrical optimal design of axially compressed cylindrical shells is given by I. N. Ginzburg and S. N. Kan [8], and for radially compressed cylindrical shells - by B. J. Hyman and A. W. Lucas Jr. [10] (optimal linearly varying thickness).

To a certain degree the optimization of arches under stability constraints (B. Budiansky, J. C. Frauenthal and J. W. Hutchinson [3]; C. H. Wu [24]) may be regarded as optimization of infinitely long cylindrical shells. Related problems may be also found in aeroelastic optimization; the basic concepts are discussed by G. Gerard [7] and by H. Ashley and S. C. McIntosh [2].

A general theory of optimization of shells with respect to their stability is lacking; following the paper by M. Życzkowski and J. Krużelecki (Proc. IUTAM Symp. Optimization in Structural Design, Warsaw 1973, publ. Springer 1975, 229-247) we present here two problems of this type: a parametrical optimization of a spherical panel and a variational optimization of a cylindrical shell subject to pure bending.

6.2 Parametrical optimal design of a spherical panel

As a simple example consider optimal design of a spherical panel under normal external pressure q, spread over a circle of a given radius a, Fig. 1. The support is assumed to realize pure membrane state before buckling takes place, thus very shallow panels will be excluded.

The objective of the design will be to minimize the volume V (or weight) of the shell. The design variable will be the central angle α, connected with the radius of the sphere:

$$R = \frac{a}{\sin \alpha} . \tag{2.1}$$

We confine the attention to the shell of a given (spherical) shape and of a constant thickness h, thus the problem belongs to parametrical optimal design. It corresponds to the design of a symmetrical two-bar truss, discussed in detail by R. Wojdanowska-Zajac and M. Zyczkowski [23]. The only constraint will be connected with the shell stability; for the large q and small "a" the strength constraint may be more important (as discussed in [23]), but this case will not be dealt with here. Elastic buckling will be assumed.

The design objective is determined by

$$V = 2\pi R^2 h(1 - \cos \alpha). \tag{2.2}$$

The simplest formulation of the constraint may be written in the form

$$q_{cr} = qj = \frac{2E}{\sqrt{3(1 - \nu^2)}} \frac{h^2}{R^2} , \tag{2.3}$$

where E and ν are the elastic constants, j = safety factor against buckling. This formula corresponds to the upper critical pressure for a complete spherical shell, but is also recommended as a first approximation for simply supported spherical panels [20,22].

Eliminating h from (2.3) and R from (2.1) we express V in terms of the angle α and of the given radius a,

$$V = \pi a^3 \sqrt{\frac{2qj}{E} \sqrt{3(1 - \nu^2)}} \frac{1 - \cos \alpha}{\sin^3 \alpha} . \tag{2.4}$$

The optimization condition $\partial V/\partial \alpha = 0$ leads to the equation

$$2 \cos^2 \alpha - 3 \cos \alpha + 1 = 0. \tag{2.5}$$

Hence $\alpha_{opt} = 60°$, $R_{opt} = 2a/\sqrt{3}$. It may be shown that the same result is obtained if only the strength constraint is used (in contradistinction to the two-bar truss), while the corresponding thickness is given by a different formula.

More exactly, the semiempirical formula determining the critical pressure for a spherical panel is proposed by K. Klöppel and O. Jungbluth (cf. [20]).

$$q_{cr} = qj = 0.3E\left(1 - 0.175\,\frac{\alpha^\circ - 20^\circ}{20^\circ}\right)\left(1 - \frac{0.07R}{400h}\right)\left(\frac{h}{R}\right)^2, \tag{2.6}$$

which may be applied within the interval $20^\circ \leq \alpha \leq 60^\circ$. Using (2.6) instead of (2.3) and solving a quadratic equation with respect to h we find

$$V = \pi a^3\,\frac{1 - \cos\alpha}{\sin^3\alpha}\left[\frac{0.07}{400} + \sqrt{\frac{4qj}{0.3E\left(1 - 0.175\,\frac{\alpha^\circ - 20^\circ}{20^\circ}\right)} + \left(\frac{0.07}{400}\right)^2}\,\right]. \tag{2.7}$$

The optimization condition $\partial V/\partial\alpha = 0$ may be then written

$$\frac{qj}{E} = \left(\frac{0.07}{400}\right)^2 \frac{\mu^2}{\Phi^2}\left(\frac{10}{3\lambda} - \frac{\Phi}{\mu}\right), \tag{2.8}$$

where μ, λ, and Φ denote the following functions of the angle α

$$\mu = \frac{2\cos^2\alpha - 3\cos\alpha + 1}{\sin\alpha},$$

$$\lambda = 1 - 0.175\,\frac{\alpha^\circ - 20^\circ}{20^\circ}, \tag{2.9}$$

$$\Phi = \frac{20}{3}\,\frac{\mu}{\lambda} + 1.672\,\frac{1 - \cos\alpha}{\lambda^2}.$$

Equation (2.8) constitutes therefore a transcendental equation with respect to the optimal angle $\alpha = \alpha_{opt}$. For a relative wide range of practical interest $10^{-6} < qj/E < 10^{-3}$ we find $\alpha_{opt} \approx 51^\circ$, thus the optimal angle is smaller than that determined by the simplest formula (2.3).

6.3 Optimal design of a cylindrical shell under pure bending

6.3.1 Concept of a "shell of uniform stability"

Consider a cylindrical shell subject to end couples M, resulting to pure bending of the shell as a whole, Fig. 2. The design objective-minimal volume of the shell - will be expressed here by the cross-sectional area A, if we consider the shell as a closed thin-walled beam. The considered design variables are, in general, the shape of the middle surface y = y(x) and the (variable) wall thickness $h = h(x)$. Such an optimization problem was stated in [26] under the constraints connected with strength, elastic stiffness and stability, but the solutions were obtained with only two first constraints taken into account. The stability constraint was then considered as an additional one applied for the determination of certain parameters. A similar approach was used in the paper by S. Mazurkiewicz and M. Zyczkowski [14] devoted to a thin-walled bar (cylindrical shell) under torsion with bending. A. Mioduchowski and K. Thermann [16] solved the problem of pure bending under the constraint of lateral buckling. Local buckling was taken into account in an early paper by B. Herbert [9], who presented a simple parametrical optimization confined to a circular cylindrical shell of constant thickness.

In the present paper, we are interested, above all, in the constraint of local wall stability. The strength condition will be also introduced (limitation of the upper bound of stresses). Neither the lateral buckling nor the Brazier effect will be taken into account. The interaction of the Brazier effect and of the local buckling was discussed in the paper by E. L. Akselrad [1]; for elastic circular cylindrical shells this effect was found to be less important than the local buckling, as a rule.

Suppose that a certain part of the material is concentrated in the outer fibers; such a situation is typical for pure bending [26]. Denote the corresponding area by A_0; of course, this area may be also designed as a horizontal flange which increases the lateral stability. Thus we are looking for the minimal value of the functional (assuming bisymmetrical cross-section)

$$A = \oint h(x)ds + 2A_0 \tag{3.1}$$

under the isoperimetric condition of a given bending moment

$$M = \oint \sigma_2(x)y(x)h(x)ds + 2A_0Hk, \tag{3.2}$$

where

$$ds = (1 + y'^2)^{1/2}\, dx \tag{3.3}$$

denotes the elementary length of the arc, $2H = 2y(0)$ is the height of the cross-section, k - the admissible normal stress. Taking the strength condition into account we present the stress distribution in the form

$$\sigma_z(x) = \frac{y(x)}{H} k. \tag{3.4}$$

The design objective (3.1) and the isoperimetric condition (3.2) contain two unknown functions, $y(x)$ and $h(x)$. Now we assume a certain approximate form of the buckling constraint, joining both these functions. In the case of a non-homogeneous stress distribution some authors (e.g. A. S. Volmir [2]) propose to apply the stability condition in the local form

$$\sigma_{cr} = \sigma_z j = \beta E \frac{h}{\rho}, \tag{3.5}$$

where β is a coefficient depending on the assumed theory of buckling ($\beta = 0.606$ for the linear theory - upper critical pressure, whereas much lower values of β are assumed for the lower critical pressure), while $\rho = \rho(x)$ denotes the corresponding radius of curvature. In the present paper we assume (3.5) to be satisfied not only at the dangerous point, but at any point of the shell. Such a shell will be called "shell of uniform stability". This approach makes it possible to eliminate one of the unknown functions (design variables), namely, combining (3.4) and (3.5) we obtain the relation

$$h(x) = \frac{jK}{\beta EH}\, y(x)\rho(x) = \frac{y(1+y'^2)^{3/2}}{\psi H|y''|}\,, \tag{3.6}$$

where ψ denotes the dimensionless parameter

$$\psi = \frac{\beta E}{jk}\,. \tag{3.7}$$

The assumption of uniform stability (3.6) is, in principle, justified since this relation gives minimal admissible wall thickness and makes it possible to concentrate maximal percentage of the mass in the outer fibers. Of course, such a simplifying assumption may lead to correct results under some restrictions only. It is certainly not valid if the curvature changes its sign, thus we assume the convexity of the shell, $d^2y/dx^2 < 0$ for $y > 0$ (some exceptions will be discussed later). The assumed double symmetry of the cross-section is justified, if the sign of the bending moment may be changed.

It will be more convenient to solve the optimization problem in dual formulation, looking for the maximal bending moment M at constant cross-sectional area A. We substitute (3.3) and (3.6) into (3.1) and eliminate the concentrated area

$$A_0 = \frac{A}{2} - \frac{2}{\psi H}\int_0^a \frac{y}{(-y'')}\,(1+y'^2)^2\,dx, \tag{3.8}$$

where 2a is the width of the cross-section at the neutral axis. Combining (3.2) and (3.8) we arrive at the functional

$$M = AHk - \frac{4k}{\psi H^2} \int_0^a \frac{y}{(-y'')} (H^2 - y^2)(1 + y'^2)^2 \, dx, \tag{3.9}$$

which is to be maximized without any isoperimetric condition.

Introduce now the following dimensionless quantities, variables and parameters:

$$\xi = \frac{x}{H}, \quad \eta = \frac{y}{H}, \quad m = \frac{M}{A^{3/2}\psi^{1/2}k}, \quad \vartheta = \frac{H}{\sqrt{A\psi}}, \quad \alpha = \frac{H}{a} \tag{3.10}$$

and rewrite (3.9) in the form

$$m = \vartheta - 4\vartheta^3 \int_0^{1/\alpha} \frac{\eta}{(-\eta'')} (1 - \eta^2)(1 + \eta'^2)^2 \, d\xi. \tag{3.11}$$

The function $\eta = \eta(\xi)$ and the parameters ϑ and α are subject to optimization. At first we determine the optimal value of ϑ. Denote the (positive) value of the integral in (3.11) by $\omega/4$; it does not depend on ϑ. Thus we look for the maximal value of the function

$$m = \vartheta - \omega\vartheta^3, \tag{3.12}$$

hence

$$\vartheta_{opt} = \frac{1}{\sqrt{3\omega}},$$

$$m_{max} = \frac{2}{3\sqrt{3\omega}} = \frac{2}{3}\vartheta_{opt} = \frac{1}{\sqrt{27 \int_0^{1/\alpha} \frac{\eta}{(-\eta'')} (1 - \eta^2)(1 + \eta'^2)^2 \, d\xi}}. \tag{3.13}$$

The bending moment is therefore proportional to the (optimal) height of the cross-section. Further optimization procedure will be carried out in several variants.

6.3.2 Shell of constant thickness

As a first step consider the shell of constant thickness, less optimal than that of variable thickness, but important for the manufacturing reasons and in view of some applications. Then the function $\eta(\xi)$ in (3.13) is no more arbitrary, but is determined by (3.6) with substituted h = const. If we rewrite (3.6) in the form

$$\frac{1}{\rho} = \frac{y}{C}\ , \text{ where } C = \frac{\beta EHh}{jk}\ , \tag{3.14}$$

then the analogy with the Euler problem of finite elastic deflections of a buckled column is seen. The optimal shape of the middle surface (middle line of the cross-section) will correspond to the deflection line of the column. Some parameters of this deflection line are the only ones to be determined.

The solution of the Euler problem is well known and we quote here just some intermediate equations and the final result. The first integration of the second order differential equation (3.14) gives

$$\frac{C}{\sqrt{1 + y'^2}} = \frac{y^2}{2} + D, \tag{3.15}$$

and after the second integration

$$\sqrt{C}[2E(\Phi,k^2) - F(\Phi,k^2)] - y\sqrt{\frac{2C - 2D - y^2}{2C + 2D + y^2}} = B - x. \tag{3.16}$$

The symbols F and E denote here the elliptic integrals of the first and of the second kind respectively, further

$$\Phi = \arc\sin\left(\frac{y}{\sqrt{C - D}}\sqrt{\frac{2C}{2C + 2D + y^2}}\right),$$
$$k^2 = \frac{C - D}{2C}\ , \tag{3.17}$$

and B and D stand for the integration constants.

The boundary conditions $y(0) = H$, $y(a) = 0$ determine the constant B, namely $B = a$, and lead to the following transcendental equation with respect to the constant D:

$$\sqrt{C}[2E(\Phi_0,k^2) - F(\Phi_0,k^2)] - H\sqrt{\frac{2C - 2D - H^2}{2C + 2D + H^2}} = a, \tag{3.18}$$

where $\Phi_0 = \Phi(y = H)$. Optimal values of the remaining constants "a" and C are to be determined.

Instead of "a" and C we optimize two others, more convenient dimensionless parameters:

$$\delta = \frac{H}{\psi h} = \frac{H^2}{C} = \frac{\vartheta^2 A\psi}{C}, \quad z = \frac{1}{\sqrt{1 + y_0'^2}} = \frac{H^2 + 2D}{2C}, \tag{3.19}$$

easier for interpretation. The parameter δ determines the thickness of the shell h (since the height H is given by (3.13)), and the parameter $z = \cos\theta_0$ determines the slope at the top, which is not necessarily equal to zero. Then, making use of (3.13) and (3.18), we express other parameters in terms of δ and z:

$$C = \frac{\vartheta^2 A\psi}{\delta} = \frac{A\psi}{3\omega\delta}, \quad D = Cz - \frac{H^2}{2} = \frac{A\psi}{6\omega\delta}(2z - \delta),$$

$$a = \sqrt{\frac{A\psi}{3\omega\delta}}\left[2E(\Phi_0,k^2) - F(\Phi_0,k^2) - \sqrt{\delta\frac{1 - z}{1 + z}}\right], \tag{3.20}$$

$$\Phi_0 = \arcsin\sqrt{\frac{\delta}{(1 + z)\left(1 - z + \frac{\delta}{2}\right)}}, \quad k^2 = \frac{\delta + 2(1 - z)}{4}.$$

The equation of the middle surface (3.16) may be now written in the dimensionless form as follows

$$\xi = \frac{1}{\sqrt{\delta}} \left[2E(\Phi_0,k^2) - 2E(\Phi,k^2) - F(\Phi_0,k^2) + F(\Phi,k^2)\right] -$$

$$- \sqrt{\frac{1 - z}{1 + z}} + \eta \sqrt{\frac{2(1 - z) + \delta(1 - \eta^2)}{2(1 + z) - \delta(1 - \eta^2)}} , \tag{3.21}$$

where

$$\Phi = \text{arc sin}\, \eta \sqrt{\frac{\delta}{\left(1 - z + \frac{\delta}{2}\right)\left[(1 + z) - \frac{\delta}{2}(1 - \eta^2)\right]}} . \tag{3.22}$$

In order to evaluate δ and z we have to maximize m_{max}, (3.13), or to minimize the integral ω. Since we determined the inverse function $\xi = \xi(\eta)$, (3.21), we express this integral in the suitable form:

$$\omega = 4 \int_0^1 \frac{\eta}{(-\xi'')} (1 - \eta^2)(1 + \xi'^2)^2 \, d\eta. \tag{3.23}$$

In view of (3.14) and (3.15), this integral may be here written in a much more simple form. We have

$$\frac{-\xi''}{(1 + \xi'^2)^{3/2}} = \frac{H^2}{C} \eta = \delta\eta, \tag{3.24}$$

and hence

$$\omega = \frac{4}{\delta} \int_0^1 (1 - \eta^2) \sqrt{1 + \xi'^2} \, d\eta. \tag{3.25}$$

Finally, after substitution of ξ' determined by (3.15) and integration

$$m = \frac{\delta^{5/4}}{\sqrt{54}\left[\sqrt{\delta \frac{1 - z}{1 + z}} + \left(1 + z + \frac{\delta}{2}\right) F(\Phi_0,k^2) - 2E(\Phi_0,k^2)\right]} . \tag{3.26}$$

The admissible region of paramters δ and z is determined by $0 \leq z \leq 1$, and by the inequalities $0 \leq k^2 \leq 1$, $0 \leq \sin \Phi_0 \leq 1$. Both last inequalities lead to $\delta \leq 2(1 + z)$, and the admissible region is shown in Fig. 3. Numerical optimization done with the aid of the computer Odra 1204 gives $\delta_{opt} = 3.430$, $z_{opt} = 0.981$, thus the optimal slope at the top is determined by $\theta_0 = 11°10'$ (but it turns out that assuming $\theta_0 = 0$ we loose less than one percent of m). The resulting parameters of the optimal solution are: dimensionless bending moment $m_{max} = 0.2473$, dimesionless concentrated area $2\,A_0/A = 0.7515$, dimensionless height $\vartheta = H/\sqrt{A\psi} = 0.3705$. The corresponding shape of the cross-section of the shell as a whole is shown in Fig. 4 for $\psi = \beta E/jk = 25$; assuming $A = 100\ cm^2$ we find the thickness $h = 0.218$ cm and the height $2H = 37.05$ cm. The width "a" is here negative, but it will be positive if we design the concentrated areas as the flanges (the thickness of which should be evaluated using ~~the plate~~ strip stability condition). For $\eta = 0$, joining the two-neighbouring arcs, we arrive at a concave singular point of the cross-section; however, this singularity seems not to be dangerous, since at that point $\sigma_z = 0$.

6.3.3 Shells of given shape of the middle surface

Consider now the case of a prescribed shape of the middle surface of the shell. Then the (variable) thickness is determined by (3.6) and only some parameters are subject to optimization. certain functions $y = y(x)$ the calculations are very simple.

Assume elliptic profile of a cylindrical shell

$$y = H\sqrt{1 - \left(\frac{x}{a}\right)^2}, \tag{3.27}$$

or, in dimensionless coordinates (3.10) with interchanged variables

$$\xi = \frac{1}{\alpha}\sqrt{1 - \eta^2},$$

where $\alpha = H/a$ is subject to optimization. The dimensionless bending moment after the height optimization (3.13) is here equal

$$m = \sqrt{\frac{\alpha^3}{27(0.0761 + 0.1143\alpha^2 + 0.0951\alpha^4)}} \tag{3.29}$$

and reaches its maximal value $m_{max} = 0.3881$ for $\alpha_{opt} = 1.3132$. Further optimal parameters are $2\ A_0/A = 0.2416$, $\vartheta = H/\sqrt{A\psi} = 0.5821$. The thickness is given by

$$\frac{h}{\sqrt{A\psi}} = 0.3375(1 - 1,2491\xi^2)^{3/2}\ \frac{\sqrt{1 - 1,7245\xi^2}}{\psi}\ . \tag{3.30}$$

The corresponding shape for $\psi = 25$ is also shown in Fig. 4.

As another example we discuss the parabolic cylinder

$$y = H\left(1 - \frac{x^2}{a^2}\right), \tag{3.31}$$

or, in dimensionless form,

$$\xi = \frac{1}{\alpha}\sqrt{1 - \eta}. \tag{3.32}$$

The dimensionless bending moment (3.13) is here equal

$$m = \sqrt{\frac{\alpha^3}{27(0.2094 + 0.6597\alpha^2 + 0.6926\alpha^4)}}\ , \tag{3.33}$$

and reaches its maximal value $m_{max} = 0.2224$ for $\alpha_{opt} = 1.2416$. Further optimal parameters are $2\ A_0/A = 0.4786$, $\vartheta = H/\sqrt{A\psi} = 0.3336$. The thickness is given by

$$\frac{h}{\sqrt{A\psi}} = 0.1082(1 - 1.5416\xi^2)\ \frac{(1 + 9.5408\xi^2)^{3/2}}{\psi}. \tag{3.34}$$

The corresponding shape is less optimal (smaller height) than those previously discussed (3.21) and (3.28), see Fig. 4.

6.3.4 Variational optimization

After the additionally constrained optimization, presented in Sections 3.2 and 3.3, consider now the minimal value of the function (3.23) (it means the maximal value of the bending moment m) without any other constraint. The form (3.23), expressed by the inverse function $\xi = \xi(\eta)$ is here more convenient than (3.13), since it does not contain the dependent variable ξ in the Euler-Lagrange equation: $F_\xi = 0$. This equation, originally of the fourth order, may be therefore once integrated in a general form, and will be written:

$$F_{\xi'} - \frac{d}{d\eta} F_{\xi''} = C_1. \tag{3.35}$$

Substituting for F the integrand in (3.23), we obtain

$$\xi''' = \frac{C_1 \xi''^3}{2\eta(1 - \eta^2)(1 + \xi'^2)^2} + \frac{4\xi'\xi''}{1 + \xi'^2} + \frac{(1 - 3\eta^2)\xi''}{2\eta(1 - \eta^2)}. \tag{3.36}$$

In the vicinity of $\xi = 0$, $\eta = 1$, it is more convenient to return to ξ as the independent variable; using the formulas for the derivatives of an inverse function we rewrite then (3.36) in the form

$$\eta''' = \frac{C_1 \eta''^2}{2\eta\eta'(1 - \eta^2)(1 + \eta'^2)^2} - \frac{\eta''^2(1 - 3\eta'^2)}{\eta'(1 + \eta'^2)} + \frac{\eta'\eta''(1 - 3\eta^2)}{2\eta(1 - \eta^2)}. \tag{3.37}$$

The constant C_1 may be evaluated from the transversality condition. This condition applied to the function $\xi = \xi(\eta)$ at the point $\eta = 0$ gives

$$\left(F_{\xi'} - \frac{d}{d\eta} F_{\xi''}\right)\bigg|_{\eta=0} = 0. \tag{3.38}$$

The comparison of (3.35) for $\eta = 0$ and of (3.38) yields $C_1 = 0$, thus the corresponding terms in (3.36) and (3.37) should be dropped.

The equations (3.36) and (3.37) are integrated numerically, starting from the point $\xi = 0$, $\eta = 1$. The derivative $\eta'(0)$ was assumed to be zero, although in Section 3.2 we found that this assumption is not strictly optimal. However, the calculations are then much more simple, and the gain due to the variation of $\eta'(0)$ seems to be very small, as for the constant-thickness shell.

Starting from $\xi = 0$ it is more convenient to use (3.37) instead of (3.36), but even (3.37) shows a singularity at that point. To investigate this singularity we expand the function $\eta = \eta(\xi)$ into a power series in the vicinity of $\xi = 0$. It turns out, that under the assumption $\eta'(0) = 0$ a simple ordinary power expansion holds

$$\eta = \sum_{i=0,2,4,\ldots}^{\infty} \alpha_i \xi^i, \tag{3.39}$$

where $\alpha_0 = 1$. Substituting (3.39) into (3.37) we express the subsequent coefficients α_i, $i = 4,6,\ldots$, in terms of the coefficient α_2, which remains arbitrary (subject to optimization or to an additional constraint):

$$\alpha_4 = \frac{\alpha_2^2}{30}(32\alpha_2 + 3),$$

$$\alpha_6 = \frac{\alpha_2^3}{31500}(32\alpha_2 + 3)(3232\alpha_2 + 153) - \frac{\alpha_2^3}{140}(96\alpha_2^2 + 5) \tag{3.40}$$

$$\cdots\cdots\cdots\cdots\cdots\cdots\cdots\cdots\cdots\cdots$$

The procedure of integration was as follows. For $0 \le \xi \le 0.01$ the series (3.39) was used. Then (3.37) was integrated numerically, using the Runge-Kutta method with the step $\Delta\xi = 0.001$. However, at a certain point ξ_1 (Fig. 5) the derivative η' increases infinitely, and the integration of (3.37) is impossible. Thus starting from a certain point ξ_0, η_0 (at which

$\eta' = -1$) the equation (3.36) was employed with the step $\Delta\eta = -0{,}001$. Inside the interval of decreasing ξ the sign at y'' or η'' in (3.8), (3.9),... and consequently in (3.37) should be changed, but the signs in (3.36) remain without alteration.

Such an integration was carried out for various values of the parameter α_2 (it means, for various values of the derivative $\eta''(0)$). The integral (3.23) was evaluated. It reaches its minimal value for $\alpha_{2opt} = -1.103$. Other parameters of the optimal solution are: $\vartheta = H/\sqrt{A\psi} = 0.7537$, $2A_0/A = 0.1819$, $m_{max} = 0.5025$. Thus the maximal bending moment is here more than 100 percent higher than for the optimal constant-thickness shell, and about 30 percent higher than for the elliptic shell of variable thickness.

The thickness of the variationally optimized shell was found from (3.6). The shape of a quadrant of this shell is shown in Fig. 4 for $\psi = 25$. Moreover, Fig. 6 shows the whole shape of the shell; the thickness is here doubled in order to present this shape more clearly. The concentrated areas are designed as flanges with the thickness $h(0)$; actually the flange thickness should be larger, since the stability of a flat element is smaller than that of a curved shell.

The points $\eta = 0$ are singular concave points, and the thickness h is here equal to zero. In view of $\sigma_z = 0$ at these points the situation seems not to be dangerous; however, because of the Brazier effect and to ensure the continuity of the whole shell, some corrections are here necessary. To avoid the singularity and the concavity we may also go down with the absolute value of requiring $\eta' = \infty$ for $\eta = 0$; the corresponding numerical calculations were not carried out. Such an approach will be obviously connected with a certain decrease of the bending moment m_{max}

On the other hand, a certain further gain may be reached by admitting the variation of the shape and of the thickness along the axis of the shell; this problem cannot be solved using the concept of the "shell of uniform stability" and will not be discussed here.

Figures, Part III, 6

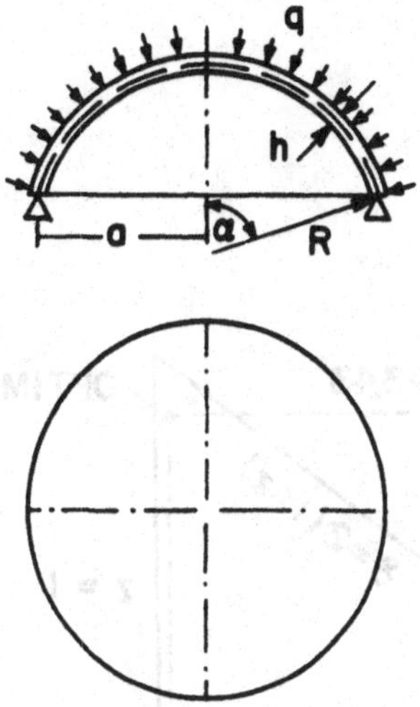

Figure 1
Parametrical optimization of a spherical panel

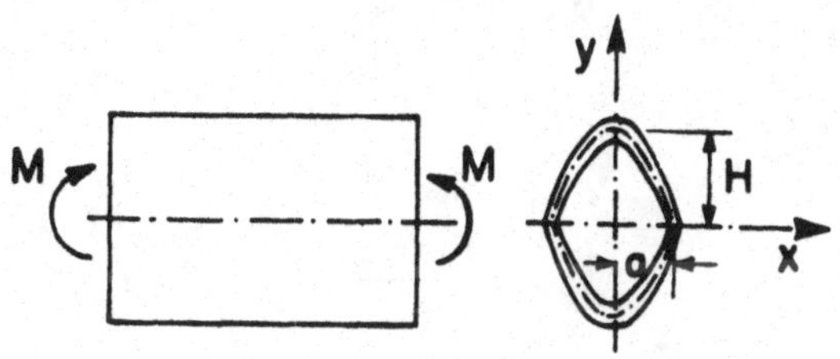

Figure 2
Noncircular cylindrical shell subject to pure bending

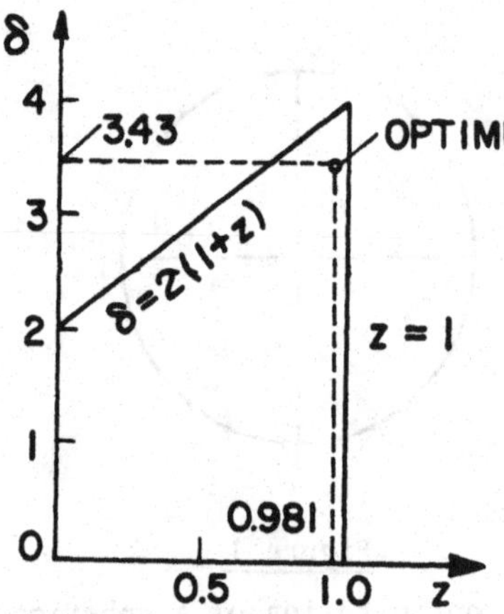

Figure 3
Admissible region of parametrical optimization

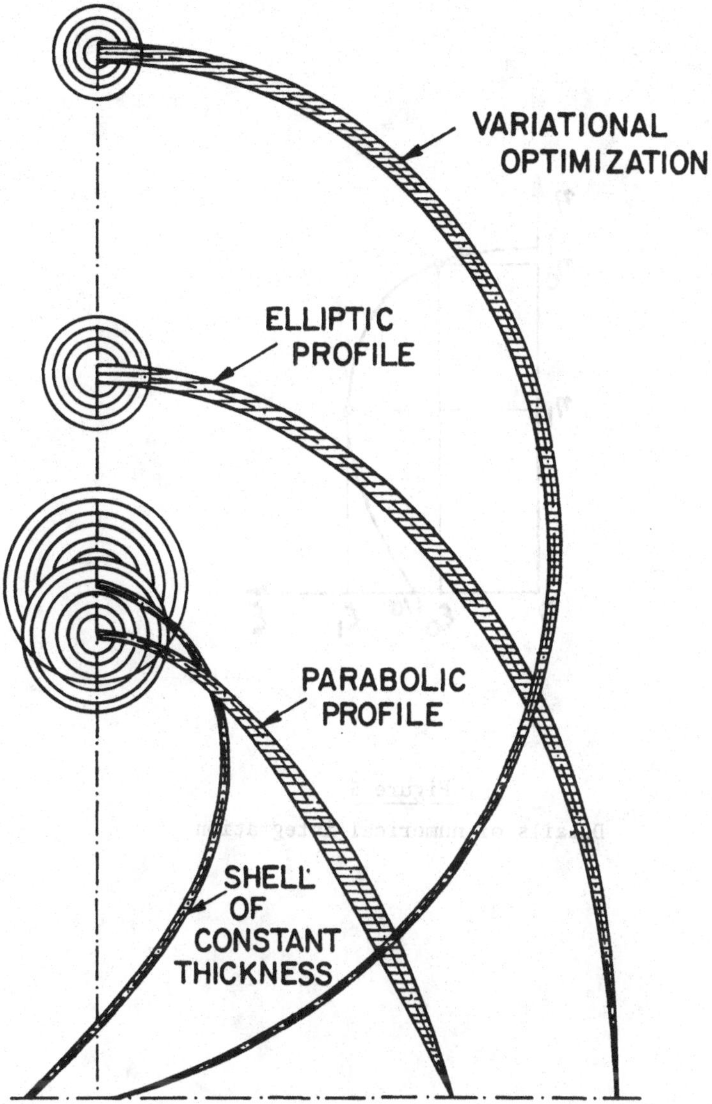

Figure 4

Optimal shell profiles under various assumptions

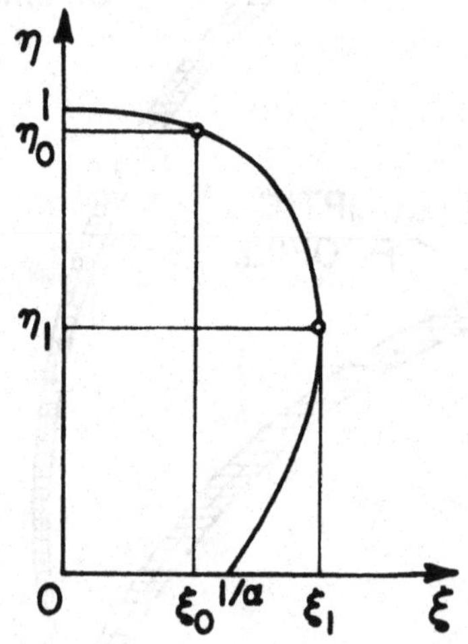

Figure 5

Details of numerical integration

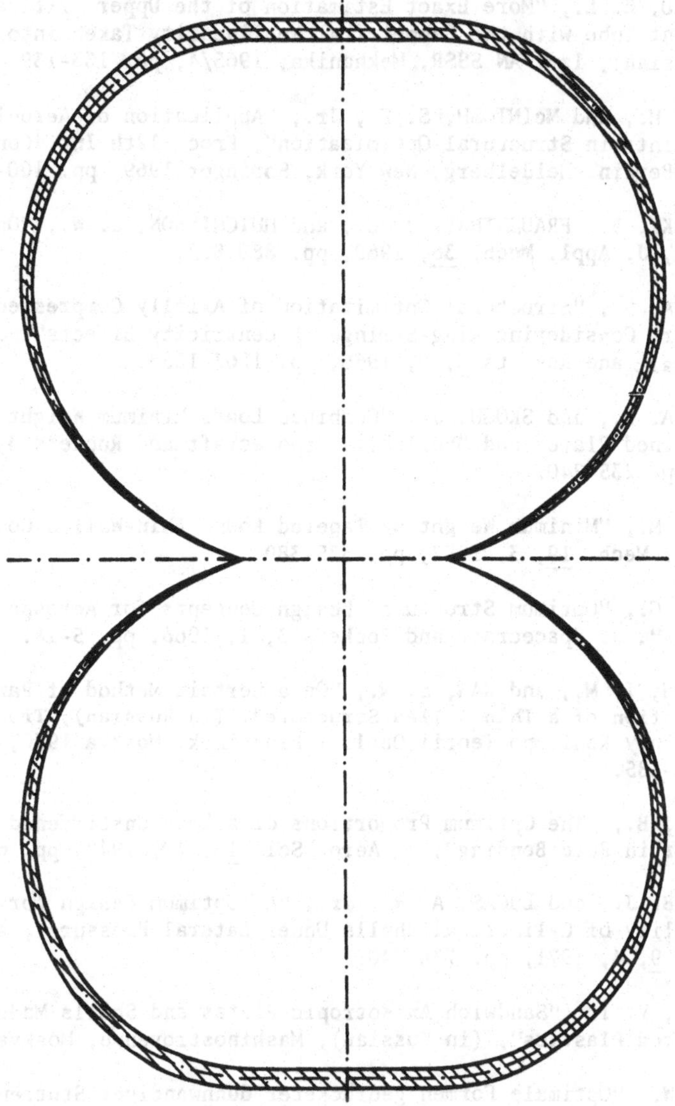

Figure 6

Optimal shell profile in variational treatment

References, Part III, 6

[1] AKSELRAD, E. L., "More Exact Estimation of the Upper Critical Load of a Bent Tube with the Geometrical Nonlinearity Taken into Account", (in Russian), Izv. AN SSSR, Mekhanika, 1965/4, pp. 133-139.

[2] ASHLEY, H., and McINTOSH, S. C., Jr., "Application of Aeroelastic Constraints in Structural Optimization", Proc. 12th Int. Congr. Appl. Mech., Berlin, Heidelberg, New York, Springer 1969, pp. 100-133.

[3] BUDIANSKY, B., FRAUENTHAL, J. C., and HUTCHINSON, J. W., "On Optimal Arches", J. Appl. Mech. 36, 1969, pp. 880-882.

[4] BURNS, A. B., "Structural Optimization of Axially Compressed Cylinders Considering Ring-Stringer Eccentricity Effects", J. Spacecraft and Rockets 3, 8, 1966, pp. 1263-1268.

[5] BURNS, A. B., and SKOGH, J., "Combined Loads Minimum Weight Analysis of Stiffned Plates and Shells", J. Spacecraft and Rockets 3, 2, 1966, pp. 235-240.

[6] FEIGEN, M., "Minimum Weight of Tapered Round Thin-Walled Columns", J. Appl. Mech. 19, 3, 1952, pp. 375-380.

[7] GERARD, G., "Optimum Structural Design Concepts for Aerospace Vehicles", J. Spacecraft and Rockets 3, 1, 1966, pp. 5-18.

[8] GINZBURG, I. M., and KAN, S. N., "On a Certain Method of Parametrical Optimization of a Thin Walled Structure", (in Russian), Trudy VII Vsesoyuznoy Konf. po Teorii Obol. i Plastinok, Moskva 1970, pp. 181-185.

[9] HERBERT, B., "The Optimum Proportions of a Long Unstiffened Circular Cylinder in Pure Bending", J. Aero. Sci. 15, 10, 1948, pp. 616-624.

[10] HYMAN, B. J., and LUCAS, A. W., Jr., "An Optimum Design for the Instability of Cylindrical Shells Under Lateral Pressure", AIAA Journal 9, 4, 1971, pp. 738-740.

[11] KOROLEV, V. I., "Sandwich Anisotropic Plates and Shells Made of Reinforced Plastics", (in Russian), Mashinostroyenye, Moskva, 1965.

[12] KRZYŚ, W., "Optimale Formen gedrückter dünnwandiger Stutzen im elastisch-plastischen Bereich", Wiss. Z. Techn. Univ., Dresden 17, 24, 1968, pp. 407-410.

[13] ŁUKASIEWICZ, S., "The Equations of the Technical Theory of Shells of Variable Rigidity", Arch. Mech. Stosowancj 13, 1, 1961, pp. 107-116.

[14] MAZURKIEWICZ, S., and ZYCZKOWSKI, M., "Optimum Design of Cross-Section of a Thin-Walled Bar Under Combined Torsion and Bending", Bull. Acad. Pol. Sci., Ser. Sci. Techn. 14, 4, 1966, pp. 273-281 (English Summary); Rozprawy Inzynierskie 14, 2, 1966, pp. 199-213, (Polish Full Text).

[15] MIODUCHOWSKI, A., and THERMANN, K., "Optimale Formen des dünnwandigen geschlossenen Querschnitts eines auf Biegung beanspruchten Balkens", Z. angew. Math. Mechanik 53, 3, 1973, pp. 193-198.

[16] PLAUT, R. H., "On the Stability and Optimal Design of Elastic Structures", in "Stability", Study No. 6, Sol. Mech. Div. Univ. of Waterloo, Waterloo, 1972, pp. 547-577.

[17] SINGER, J., and BARUCH, M., "Recent Studies on Optimization for Elastic Stability of Cylindrical and Conical Shells", Roy. Aeron. Soc. Centenary Congress, London 1966, Int. Council Aero. Sci. Paper No. 66-13, pp. 1-32.

[18] SPUNT, L., "Optimum Structural Design", Englewood Cliffs, Prentice-Hall, 1971.

[19] TEREBUSHKO, O. I., "On the Stability Analysis and Design of Reinforced Cylindrical Shells", (in Russian), Coll. "Raschet Prostranstvennykh konstrukciy", VII, Moskva, 1962, pp. 119-134.

[20] TIMOSHENKO, S. P., and GERE, J. M., "Theory of Elastic Stability", 2nd Ed., New York, McGraw-Hill, 1961.

[21] VASILEV, V. V., "Optimal Structural Design of Plates and Shells", (in Russian), Trudy VII Vsesoyuznoy Konf. po Teorii Obol. i Plastinok, Moskva, 1970, pp. 722-735.

[22] VOLMIR, A. S., "Stability of Deformable Systems", (in Russian), Moskva, Nauka, 1967.

[23] WOJDANOWSKA-ZAJAC, R., and ZYCZKOWSKI, M., "Optimal Structural Design of Trusses with the Condition of Elastic-Plastic Stability Taken into Account", Bull. Acad. Pol. Ser. Sci. Techn. 18, 9, 1970, pp. 365-374, (English Summary); Rozpr. Inz. 17, 3, 1969, pp. 347-368 (Polish Full Text).

[24] WU, C. H., "The Strongest Circular Arch - A Perturbation Solution", J. Appl. Mech. 35, 1968, pp. 476-480.

[25] ZYCZKOWSKI, M., "Optimum Design of Point-Reinforcement of Cylindrical Shells with Respect to their Stability", Arch. Mech. Stosowanej 19, 5, 1967, pp. 699-714.

[26] ZYCZKOWSKI, M., "Optimale Formen des dünnwandigen geschlossenen Querschnittes eines Balkens bei Berücksichtigung von Stabilitätsbedingungen", Z. angew. Math. Mechanik 48, 7, 1968, pp. 455-462.

[27] ZYCZKOWSKI, M., "Optymalne kształtowanie wytrzymałościowe przy uwzględnieniu warunków stateczności", Coll. "Metody optymalizacji ustrojów odkształcalnych", I. Wrocław - Warszawa - Kraków 1968, PAN, pp. 466-555.

[28] ZYCZKOWSKI, M., "Optimal Structural Design in Rheology", J. Appl. Mech. 38, 1, 1971, pp. 39-46.

OTHER CISM – SPRINGER–VERLAG BOOKS ON RELATED FIELDS

No.	Author	Title
54	Lippmann	Extremum and variational Principles in Mechanics
58	H. Parkus	Variational Principles in Thermo-and Magneto-Elasticity
121	E.R. de Arantes e Olivera	Foundations of the Mathematical Theory of Structures
126	B. Fraeijs de Veubeke M. Gerardin, A. Huck M.A. Hogge	Heat Conduction – Structural Dynamics
151	W. Nowacki - W. Olszak Editors	Micropolar Elasticity
175	S. Rinaldi Editor	Topics in Combinatorial Optimization
203	D.L. Dean	Discrete Field Analysis of Structural Systems
212	W. Prager	Introduction to Structural Optimization
236	M. Esslinger, B. Geier	Postbuckling Behaviour of Structures

Printed in the United States
By Bookmasters